NHM 3
New Heinemann Maths

Teaching Resource Book

Heinemann

Heinemann Educational Publishers
Halley Court, Jordan Hill, Oxford, OX2 8EJ
a division of Reed Educational and Professional Publishing Ltd

Heinemann is a registered trademark of Reed Educational and Professional Publishing Ltd

© Scottish Primary Mathematics Group 2000

Writing team
John T Blair
Percy W Farren
Myra A Pearson
Dorothy S Simpson
John W Thayers
David K Thomson

The material in this publication is copyright. The duplicating masters (Pupil Sheets, Home Activities and Resource Sheets) may be photocopied for one-time use as instructional material in a classroom by a teacher, but they may not be copied in unlimited quantities, kept on behalf of others, passed or sold on to third parties, or stored for future use in a retrieval system. If you wish to use the material in any way other than that specified you must apply in writing to the publisher.

First published 2000

05 04 03 02 01 00
10 9 8 7 6 5 4 3 2

ISBN 0 435 17211 5

Designed and illustrated by Gecko Limited, Bicester, Oxon.
Printed and bound in Great Britain by Page Bros.

Contents

Pupil Sheets *4*

Home Activities *61*

Resource Sheets *94*

Pupil Sheets

Numbers to 1000

Pupil Sheet 1	Sequences, before, after, between, one/two more/less
Pupil Sheet 2	Sequences, before, after, between, one/two more/less
Pupil Sheet 3	Domino template
Pupil Sheet 4	Number names, multiples of 10
Pupil Sheet 5	Number names, multiples of 100

Addition to 100

Pupil Sheet 6	Adding a two-digit number and a multiple of 10
Pupil Sheet 7	Adding two-digit numbers
Pupil Sheet 8	Adding two-digit numbers
Pupil Sheet 9	Adding a two-digit number and a single-digit/teens number, with bridging
Pupil Sheet 10	Adding two-digit numbers, with bridging
Pupil Sheet 11	Adding two-digit numbers, with bridging

Subtraction to 100

Pupil Sheet 12	Subtraction facts for 20
Pupil Sheet 13	Subtraction facts to 20
Pupil Sheet 14	A single-digit number from a two-digit number
Pupil Sheet 15	Subtracting multiples of 10
Pupil Sheet 16	Subtracting multiples of 10
Pupil Sheet 17	Two-digit number from a two-digit number
Pupil Sheet 18	A teens number from a two-digit number, bridging a multiple of ten
Pupil Sheet 19	A two-digit number from a two-digit number, bridging a multiple of ten
Pupil Sheet 20	Subtraction of two-digit numbers, no bridging
Pupil Sheet 21	Exchanging 1 ten for 10 units
Pupil Sheet 22	Subtraction of two-digit numbers, with bridging

Multiplication

Pupil Sheet 23	The 2, 3, 4, 5 and 10 times tables

Division

Pupil Sheet 24	Dividing by 5
Pupil Sheet 25	Grouping with remainders
Pupil Sheet 26	Sharing with remainders

Money

Pupil Sheet 27   Counting coin collections with £2
Pupil Sheet 28   Change from £1 and £2

Fractions

Pupil Sheet 29   Tenths
Pupil Sheet 30   Tenths
Pupil Sheet 31   Half of, quarter of, tenth of
Pupil Sheet 32   Thirds
Pupil Sheet 33   Tenths, thirds, fifths
Pupil Sheet 34   Tenths, thirds, fifths
Pupil Sheet 35   Third, fifth of a set

Addition and Subtraction to 1000

Pupil Sheet 36   Adding multiples of 10
Pupil Sheet 37   Adding two-digit numbers, bridging 100
Pupil Sheet 38   Adding two-/three-digit numbers and multiples of 10
Pupil Sheet 39   Subtracting a multiple of 10 from a three-digit number
Pupil Sheet 40   Finding small differences

Capacity

Pupil Sheet 41   Millilitres

Time

Pupil Sheet 42   The calendar
Pupil Sheet 43   Writing times
Pupil Sheet 44   Minutes past and to (digital)
Pupil Sheet 45   Minutes/hours before/after

2D Shape

Pupil Sheet 46   Quadrilaterals, hexagons, octagons
Pupil Sheet 47   Strip patterns
Pupil Sheet 48   Using and applying
Pupil Sheet 49   Grid references
Pupil Sheet 50   Angles
Pupil Sheet 51   Position and movement
Pupil Sheet 52   Symmetry

Data Handling

Pupil Sheet 53   Carroll diagram
Pupil Sheet 54   Venn diagrams

Name: _____

Pupil Sheet 1

| 168 | 284 | 193 | 500 | 612 | 785 |

| 826 | 493 | 721 | 998 | 125 | 502 |

| 689 | 647 | 682 | 674 | 651 | 628 |

| 141 | 185 | 168 | 190 | 153 | 172 |

164	850	476
583	297	965
719	302	644

1000	501	727
322	916	449
688	173	845

Numbers to 1000 • *Sequences, before, after, between, one/two more/less*

New Heinemann Maths 3 © SPMG 2000
Copying permitted for purchasing school only. This material is not copyright free.

Name: _____

Pupil Sheet 2

Numbers to 1000 • *Sequences, before, after, between, one/two more/less*

Name:

Pupil Sheet 3

Domino Template

New Heinemann Maths 3 © SPMG 2000
Copying permitted for purchasing school only. This material is not copyright free.

Name: _____

Pupil Sheet 4

Write each number.

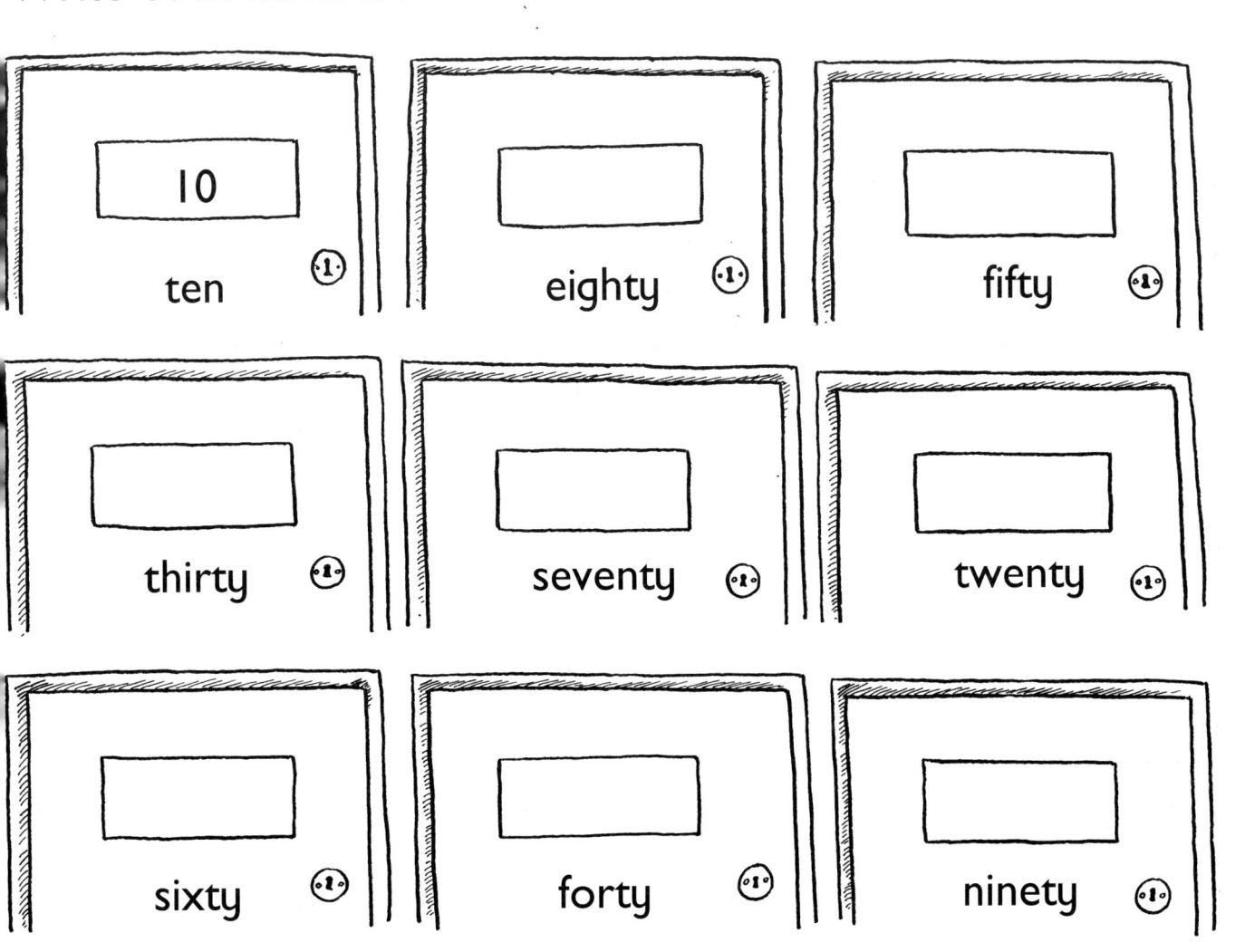

10		
ten	eighty	fifty
thirty	seventy	twenty
sixty	forty	ninety

Write each number name.

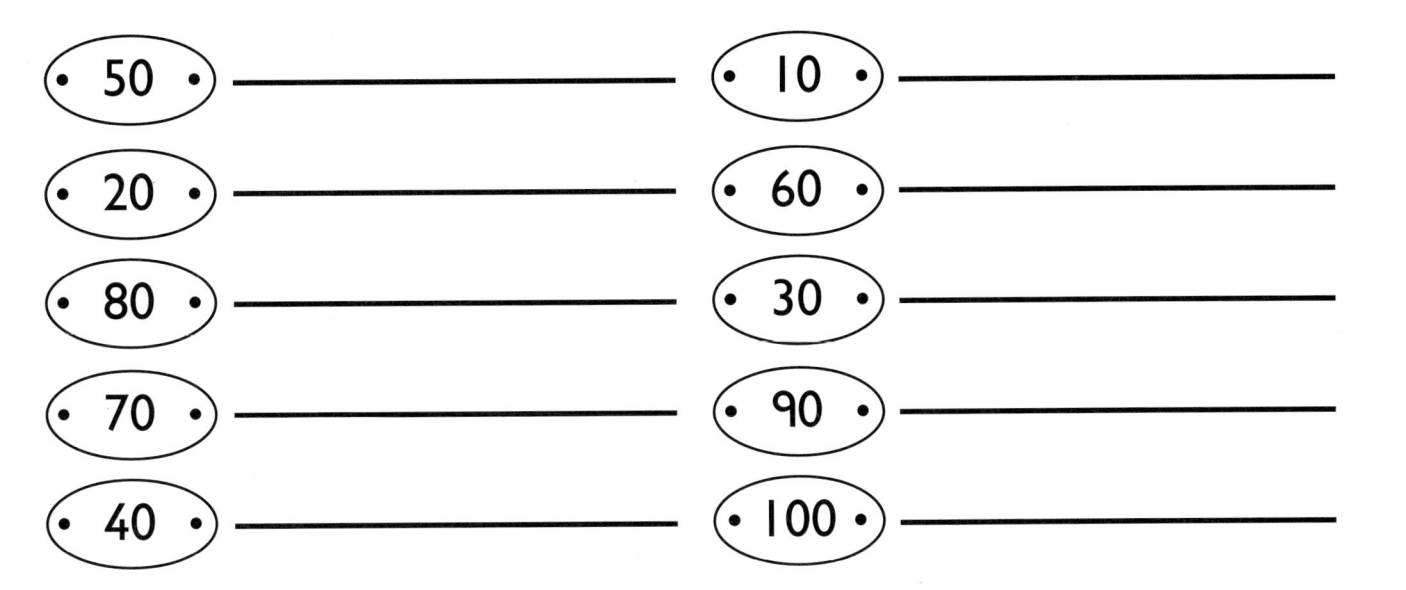

- 50 — _____
- 20 — _____
- 80 — _____
- 70 — _____
- 40 — _____

- 10 — _____
- 60 — _____
- 30 — _____
- 90 — _____
- 100 — _____

Numbers to 1000 • *Number names, multiples of 10*

Name:

Pupil Sheet 5

Write each number.

100		
one hundred	six hundred	four hundred
nine hundred	two hundred	seven hundred
three hundred	eight hundred	five hundred

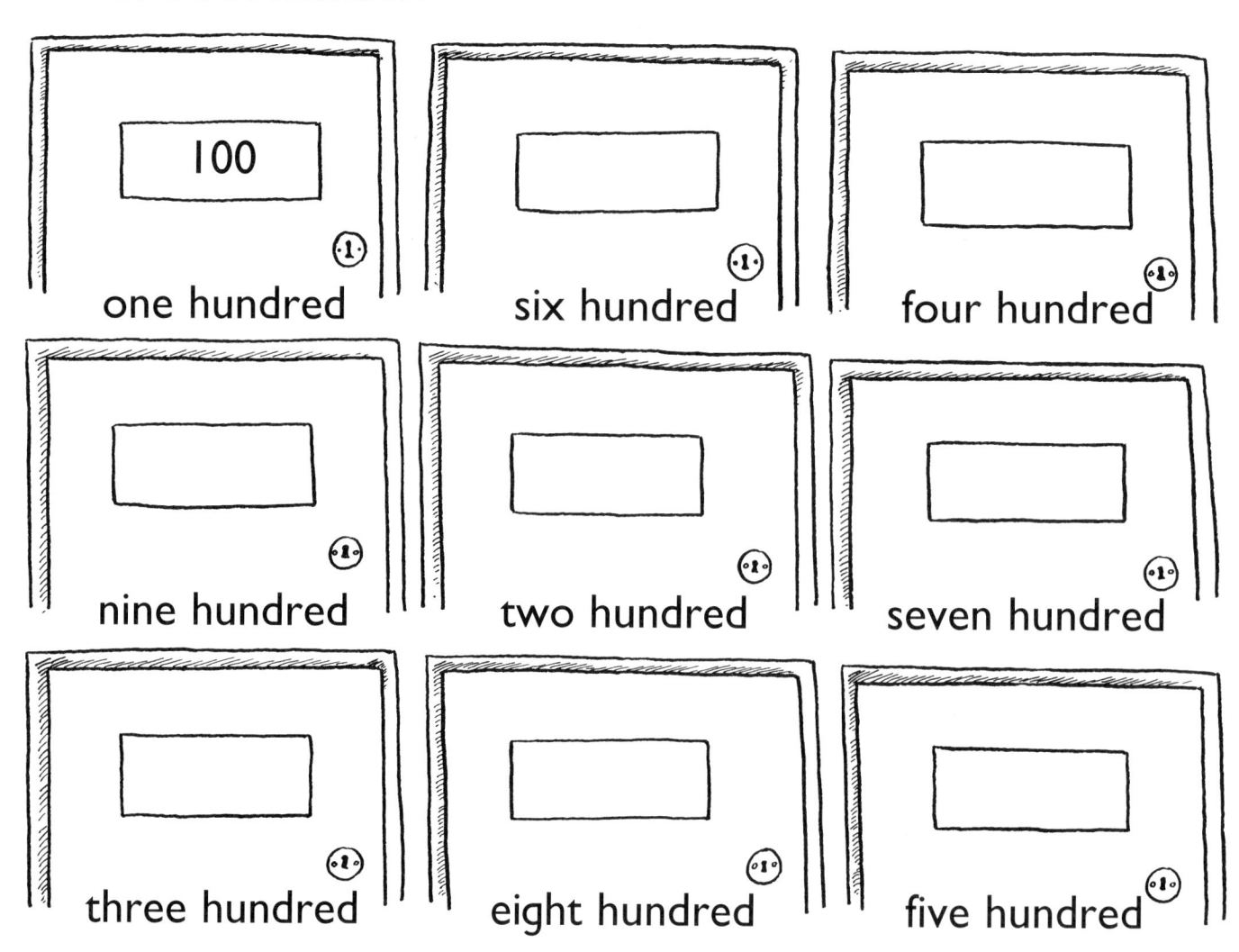

Write each number name.

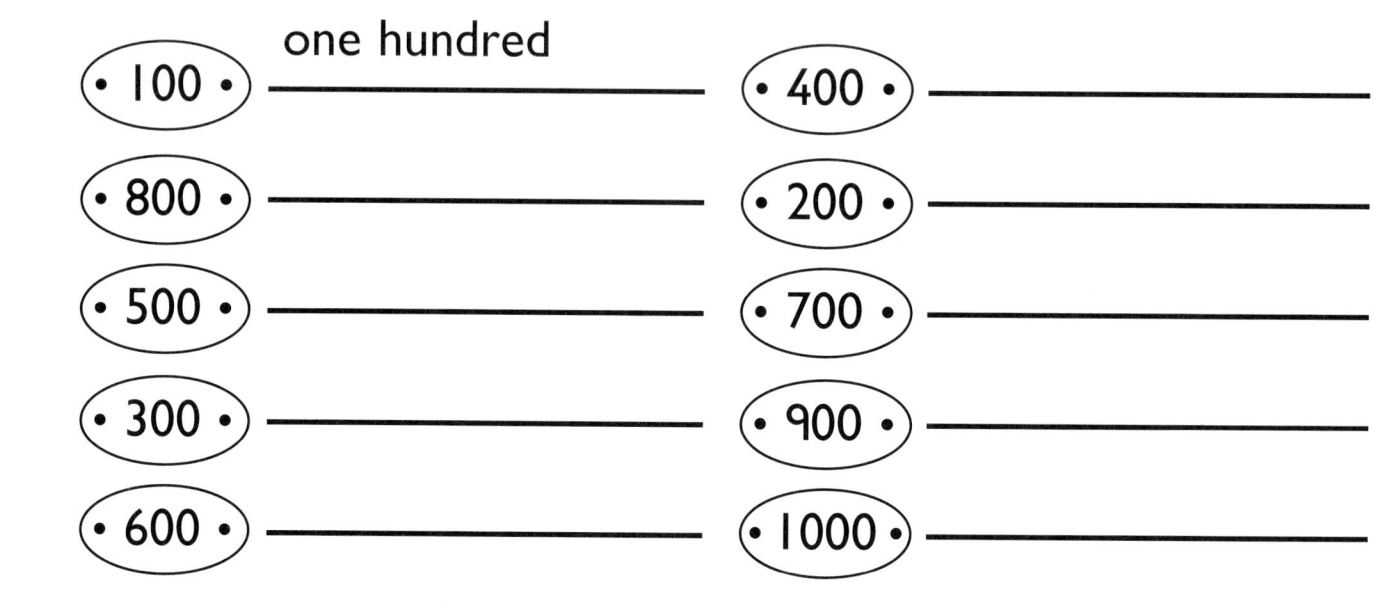

- 100 — one hundred
- 800 — _____
- 500 — _____
- 300 — _____
- 600 — _____

- 400 — _____
- 200 — _____
- 700 — _____
- 900 — _____
- 1000 — _____

Numbers to 1000 • *Number names, multiples of 100*

Name:

Pupil Sheet 6

```
            70 + 20 =
           /         \
  73 + 20 =          70 + 25 =
  78 + 20 =          70 + 22 =
  76 + 20 =          70 + 29 =


            30 + 40 =
           /         \
  34 + 40 =          30 + 48 =
  37 + 40 =          30 + 46 =
  31 + 40 =          30 + 43 =
```

20 + 50 =	60 + 30 =
29 + 50 =	60 + 32 =

10 + 70 =	40 + 40 =
15 + 70 =	40 + 47 =

Addition to 100 • *Adding a two-digit number and a multiple of 10*

Name:

Pupil Sheet 7

67 + 22 = ☐

```
      +20
    ⌒
   •_____
   67
```

25 + 63 = ☐

44 + 33 = ☐

22 + 65 = ☐

55 + 24 = ☐

32 + 37 = ☐

56 + 24 = ☐

Addition to 100 • *Adding two-digit numbers*

Name: _____

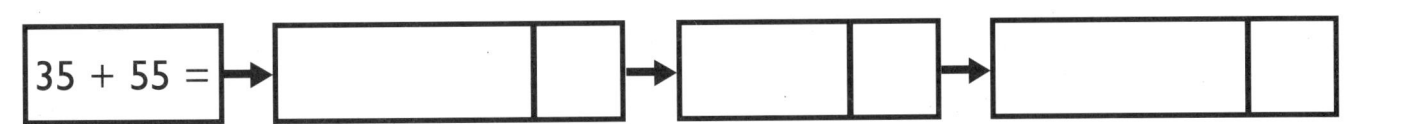

Pupil Sheet 8

| 45 + 52 = | → | 40 + 50 = | 90 | → | 5 + 2 = | 7 | → | 90 + 7 = | 97 |

26 + 73 = →

24 + 42 = →

22 + 77 = →

43 + 25 = →

64 + 34 = →

35 + 55 = →

Addition to 100 • *Adding two-digit numbers*

Name:

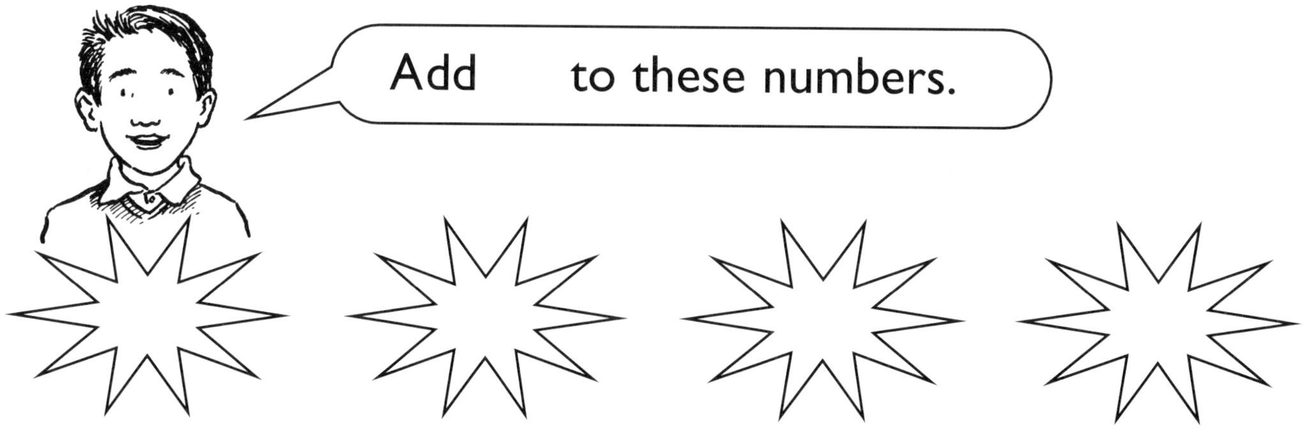

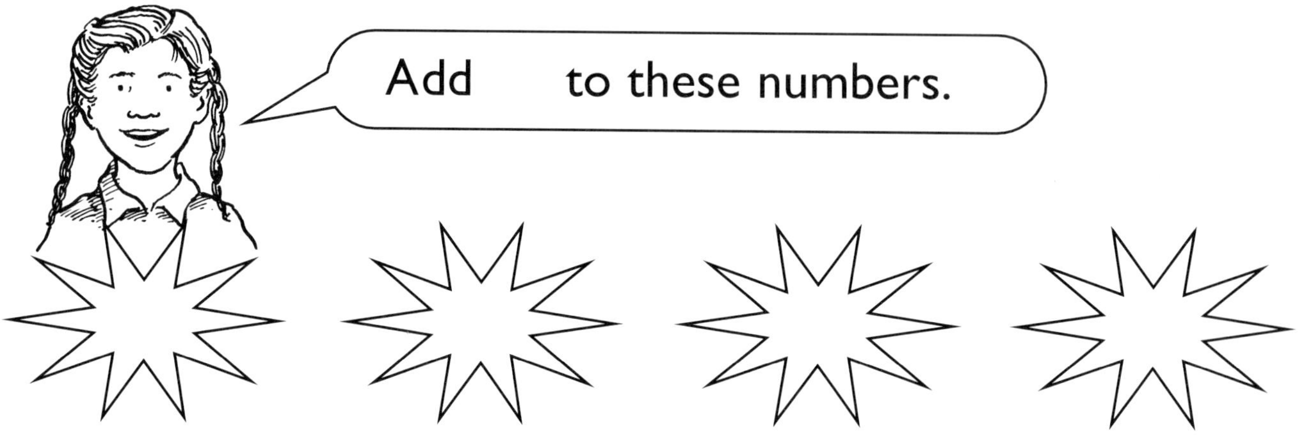

Pupil Sheet 9

Add ____ to these numbers.

☆ ☆ ☆ ☆
___ ___ ___ ___

Add ____ to these numbers.

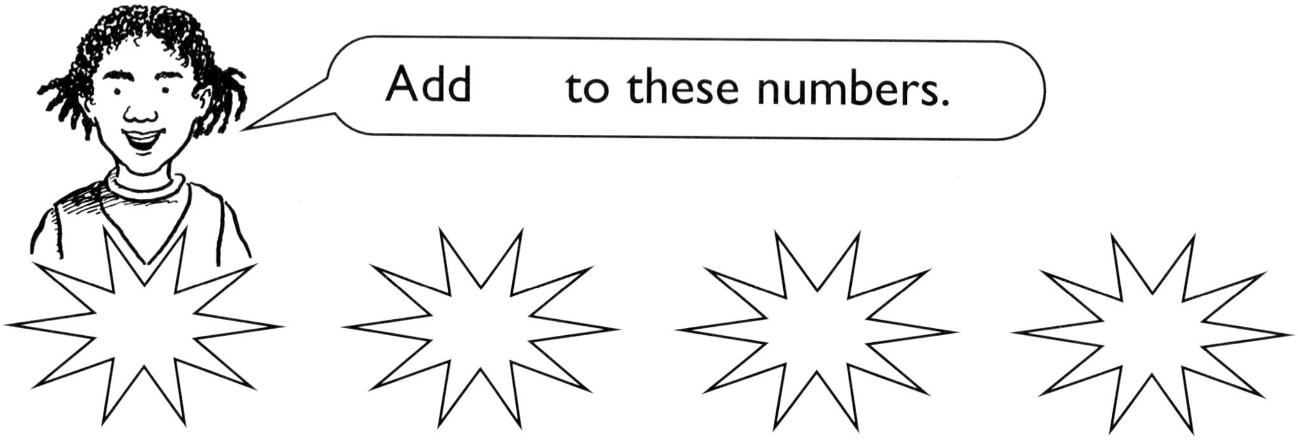

☆ ☆ ☆ ☆
___ ___ ___ ___

Add ____ to these numbers.

☆ ☆ ☆ ☆
___ ___ ___ ___

Addition to 100 • *Adding a two-digit number and a single-digit/teens number, with bridging*

New Heinemann Maths 3 © SPMG 2000
Copying permitted for purchasing school only. This material is not copyright free.

Name:

Pupil Sheet 10

57 + 24 = ☐   •―――――――――――――――
                    57

66 + 27 = ☐   •―――――――――――――――

38 + 53 = ☐   •―――――――――――――――

28 + 45 = ☐   •―――――――――――――――

47 + ☐ = 93   •―――――――――――――――

34 + ☐ = 82   •―――――――――――――――

Addition to 100 • *Adding two-digit numbers, with bridging*

New Heinemann Maths 3 © SPMG 2000
Copying permitted for purchasing school only. This material is not copyright free.

Name:

Pupil Sheet 11

$$26 + 56 \longrightarrow \begin{array}{r} 20 + 6 \\ 50 + 6 \\ \hline 70 + 12 \end{array} \longrightarrow 82$$

55 + 38 ⟶

47 + 48 ⟶

59 + 33 ⟶

27 + 69 ⟶

Addition to 100 • *Adding two-digit numbers, with bridging*

Name:

Pupil Sheet **12**

6	20 − 17
9	20 − 12
19	20 − 8
3	20 − 5
8	20 − 6
12	20 − 14
15	20 − 11
14	20 − 1

Subtraction to 100 • *Subtraction facts for 20*

New Heinemann Maths 3 © SPMG 2000
Copying permitted for purchasing school only. This material is not copyright free.

Name:

Pupil Sheet 13

Roll a 🎲. Put the number in the hexagon.
Write subtraction stories.

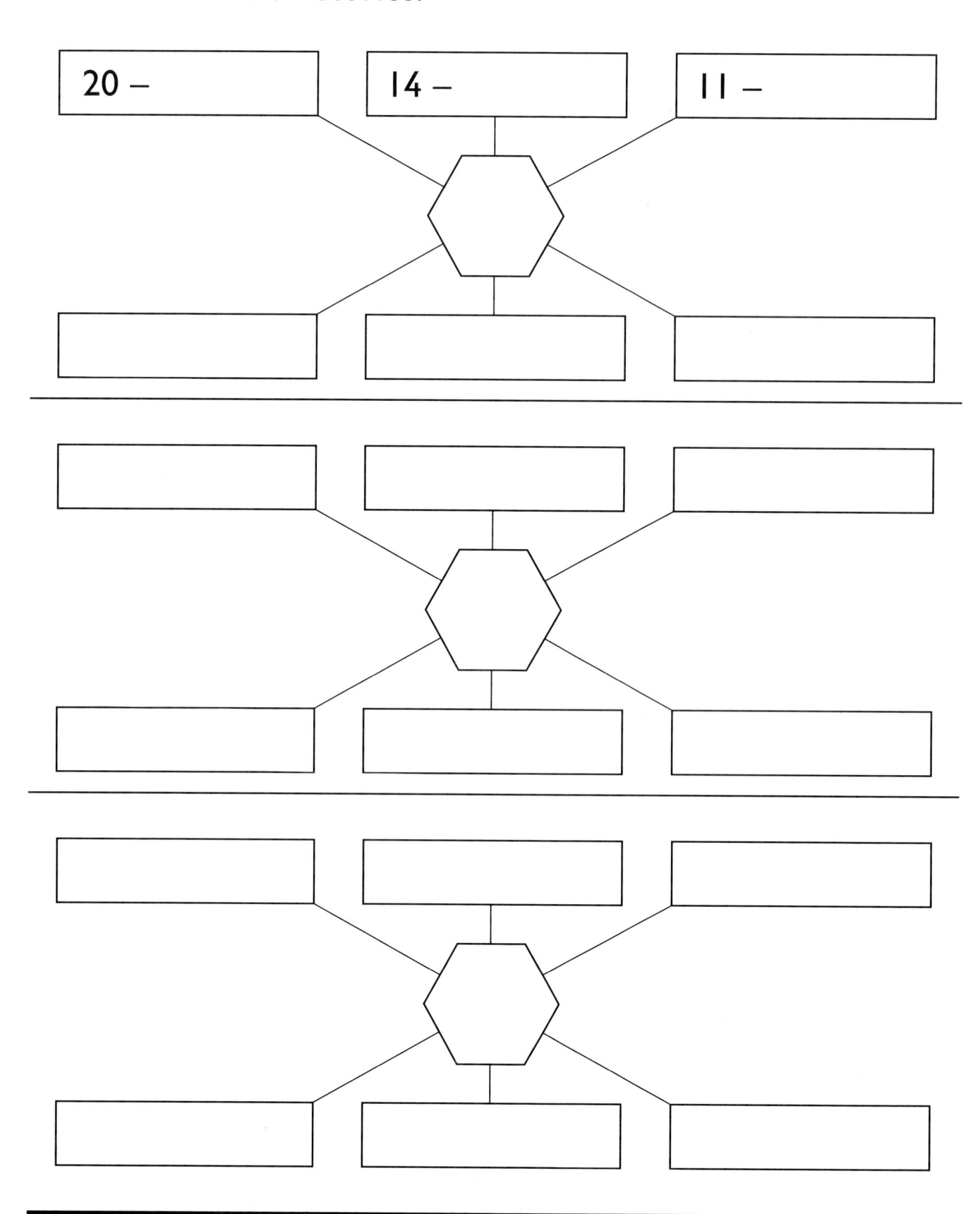

20 −   14 −   11 −

Subtraction to 100 • Subtraction facts to 20

Name: _____

Pupil Sheet 14

34 − 3 = ☐   26 − 4 = ☐

68 − 1 = ☐   89 − 8 = ☐

47 − 5 = ☐   58 − 6 = ☐

76 − 2 = ☐   55 − 4 = ☐

28 − 7 = ☐   99 − 9 = ☐

Subtract 6.
28 →
49 →
67 →
86 →

Subtract 4.
35 →
97 →
58 →
76 →

Subtract 2.
85 →
63 →
77 →
54 →

5 less than 47 ___   3 less than 86 ___

7 less than 58 ___   9 less than 79 ___

Subtraction to 100 • A single-digit number from a two-digit number

New Heinemann Maths 3 © SPMG 2000
Copying permitted for purchasing school only. This material is not copyright free.

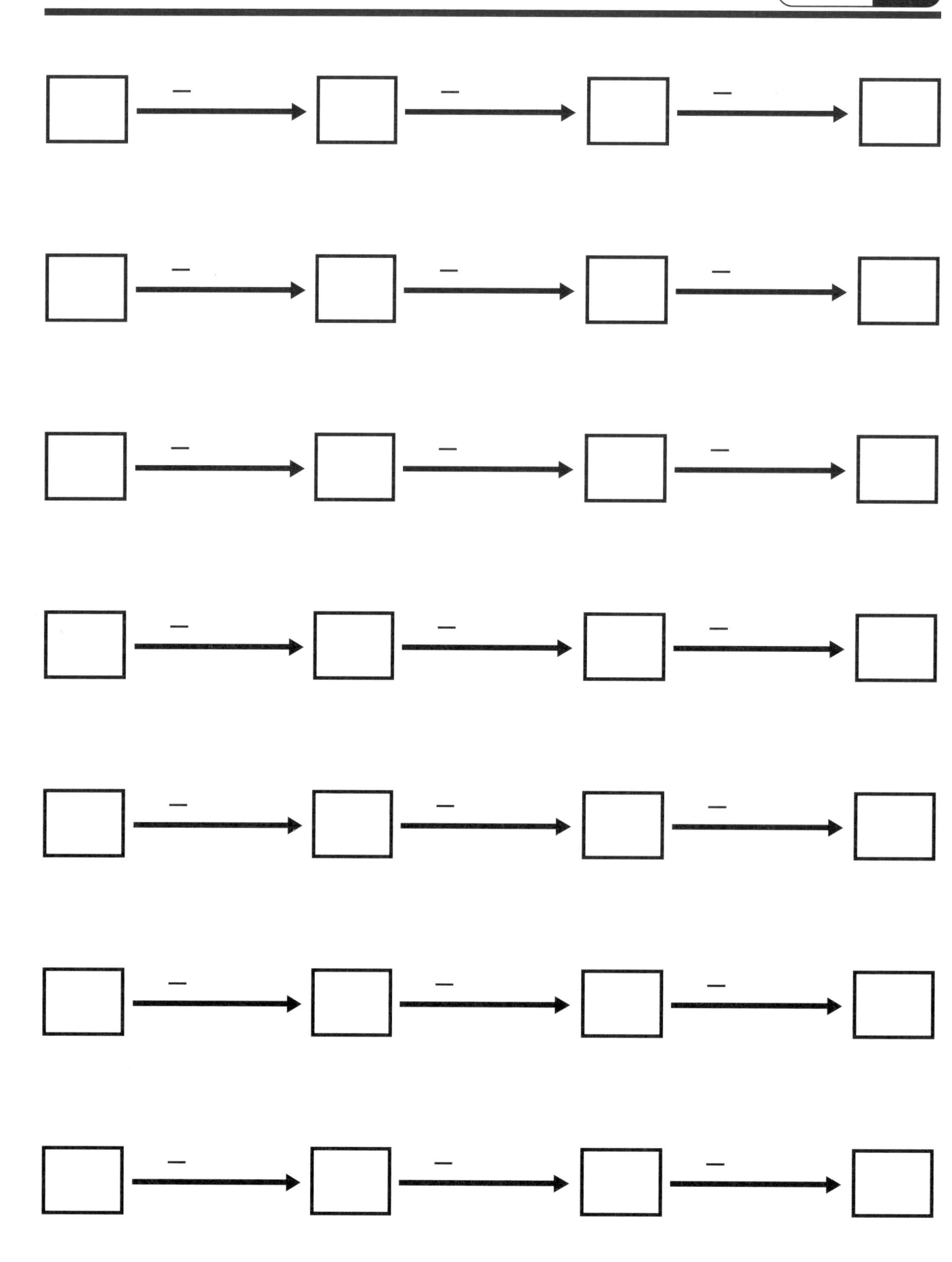

Name:

Pupil Sheet 15

Subtraction to 100 • *Subtracting multiples of 10*

Name:

Pupil Sheet 16

Cube showing: 60, 20, 40

| 98 | − | 60 | = | 38 |

☐ − ☐ = ☐   ☐ − ☐ = ☐

☐ − ☐ = ☐   ☐ − ☐ = ☐

☐ − ☐ = ☐   ☐ − ☐ = ☐

☐ − ☐ = ☐   ☐ − ☐ = ☐

Find the difference.

☐   ☐

Subtraction to 100 • *Subtracting multiples of 10*

Name: _____

Pupil Sheet 17

Jane buys 26 sunflower plants.
14 grow.
How many plants **do not** grow? ___

Tony buys 45 daises.
21 grow.
How many daises **do not** grow? ___

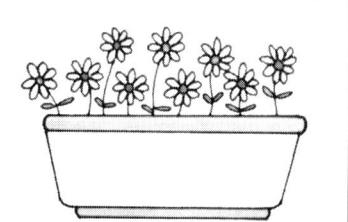

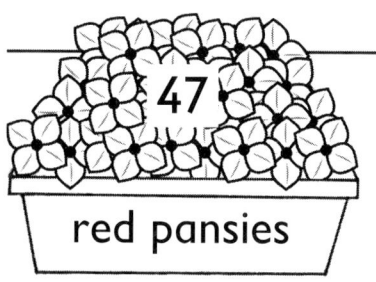

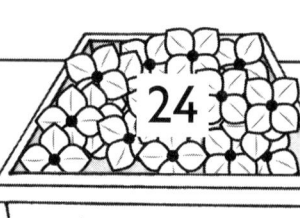

| 47 red pansies | 36 blue pansies | 13 yellow pansies | 24 white pansies |

How many more

  red pansies than yellow pansies ___

  blue pansies than white pansies ___

  blue pansies than yellow pansies ___

  red pansies than white pansies? ___

What is the difference between the number of blue and the number of red pansies? ___

74 − ☐ = 51   87 − ☐ = 45   59 − ☐ = 35

Subtraction to 100 • Two-digit number from a two-digit number

Name:

Pupil Sheet 18

33 − 16 = ☐ ──────────────●
 33

43 − 15 = ☐ ──────────────●

84 − 17 = ☐ ──────────────●

61 − 13 = ☐ ──────────────●

52 − 18 = ☐ ──────────────●

95 − 19 = ☐ ──────────────●

71 − 12 = ☐ ──────────────●

Subtraction to 100 • *A teens number from a two-digit number, bridging a multiple of ten*

Name:

Pupil Sheet 19

44 − 27 = ☐ ─────────────●
⠀⠀⠀⠀⠀⠀⠀⠀⠀⠀⠀⠀⠀⠀⠀⠀⠀⠀⠀⠀⠀⠀⠀⠀⠀⠀⠀⠀⠀⠀44

55 − 28 = ☐ ─────────────●

83 − 55 = ☐ ─────────────●

62 − 36 = ☐ ─────────────●

96 − 59 = ☐ ─────────────●

75 − 24 = ☐ ─────────────●

52 − 37 = ☐ ─────────────●

Subtraction to 100 • *A two-digit number from a two-digit number, bridging a multiple of ten*

Pupil Sheet 20

66 − 13 ⟶   6 6 ⟶ 60 + 6
            − 1 3 ⟶ 10 + 3
            ───   50 + 3 = 53

---

58 − 25 ⟶   5 8 ⟶ 50 +
            − 2 5 ⟶ 20 +
            ───   +   =

---

97 − 54 ⟶   9 7 ⟶   + 7
            − 5 4 ⟶   + 4
            ───   +   =

---

38 − 26 ⟶   3 8 ⟶   +
            − 2 6 ⟶   +
            ───   +   =

---

88 − 56 ⟶   8 8 ⟶   +
            − 5 6 ⟶   +
            ───   +   =

Subtraction to 100 • *Subtraction of two-digit numbers, no bridging*

Name:

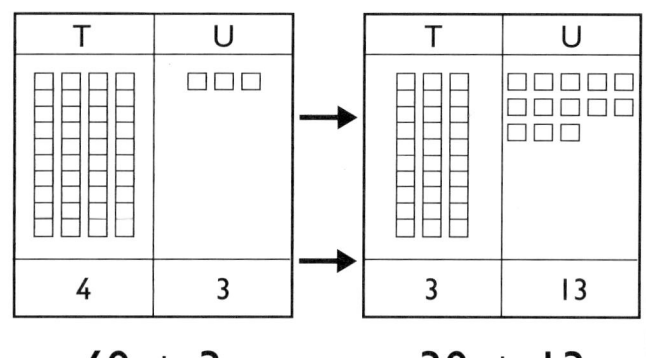

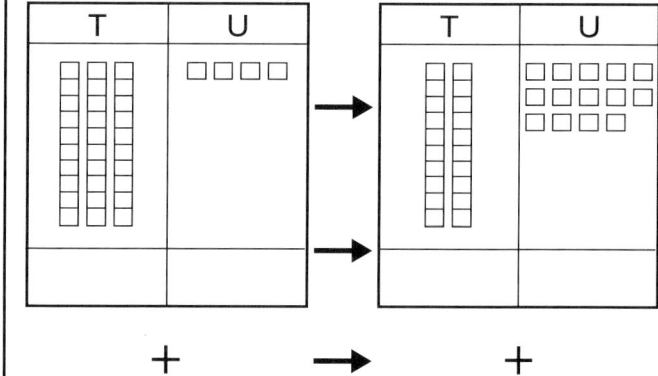

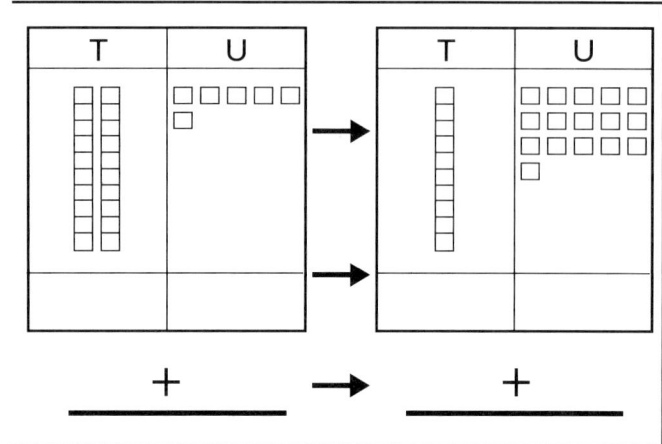

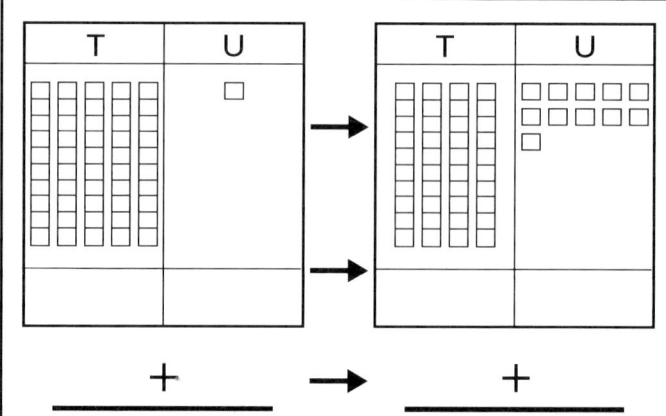

Pupil Sheet 21

T	U		T	U
4	3	→	3	13

40 + 3 → 30 + 13

T	U		T	U

___ + ___ → ___ + ___

T	U		T	U

___ + ___ → ___ + ___

T	U		T	U

___ + ___ → ___ + ___

4 7 → 40 + 7 → 30 + 17

8 3 → ___ + ___ → ___ + ___

6 5 → _____ → _____

9 8 → _____ → _____

7 2 → _____ → _____

Subtraction to 100 • *Exchanging 1 ten for 10 units*

Name:

Pupil Sheet **22**

63 − 16 ⟶   6 3 ⟶ 60 + 3 ⟶ 50 +
           − 1 6 ⟶ _____ + _____ ⟶ 10 + _____
                                                                = _____

72 − 27 ⟶   7 2 ⟶ _____ + _____ ⟶ _____ + _____
           − 2 7 ⟶ _____ + _____ ⟶ _____ + _____
                                                                = _____

54 − 36 ⟶   5 4 ⟶ _____ + _____ ⟶ _____ + _____
           − 3 6 ⟶ _____ + _____ ⟶ _____ + _____
                                                                = _____

93 − 58 ⟶   9 3 ⟶ _____ + _____ ⟶ _____ + _____
           − 5 8 ⟶ _____ + _____ ⟶ _____ + _____
                                                                = _____

85 − 49 ⟶   8 5 ⟶ _____ + _____ ⟶ _____ + _____
           − 4 9 ⟶ _____ + _____ ⟶ _____ + _____
                                                                = _____

Subtraction to 100 • *Subtraction of two-digit numbers, with bridging*

Pupil Sheet 23

Name:

×	10	3	5	1	0	2	4
9							
1							
5							
3							
10							
2							
7							
4							
8							
0							
6							

Multiplication • *The 2, 3, 4, 5 and 10 times tables*

Pupil Sheet 24

Name:

| 1 | 45 divided by 5 |
| 6 | How many fives are in 5? |

| 10 | Divide 25 equally among 5. |
| 3 | Divide 35 by 5. |

| 9 | Share 15 equally among 5. |
| 2 | 50 divided by 5 |

| 5 | Divide 30 by 5 |
| 7 | How many fives are in 10? |

Division • Dividing by 5

Name:

Pupil Sheet 25

Use cubes.

Ring groups of 2.

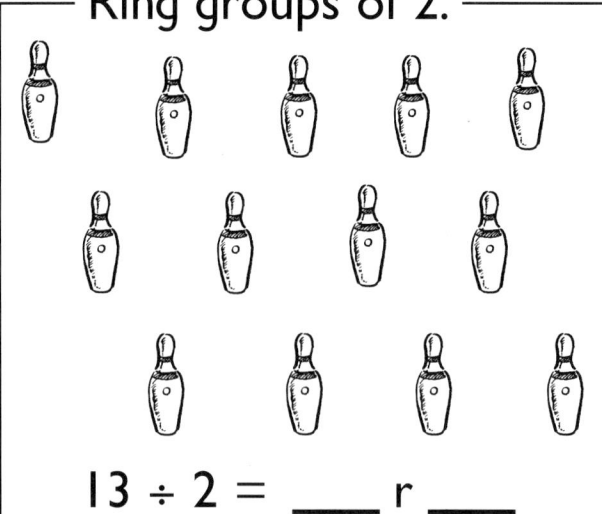

13 ÷ 2 = ___ r ___

Ring groups of 3.

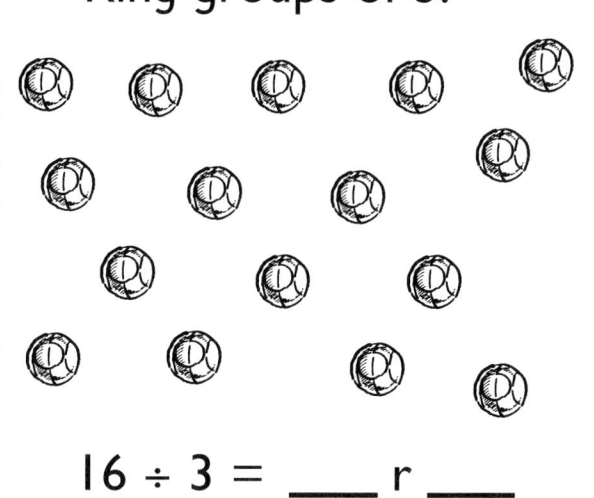

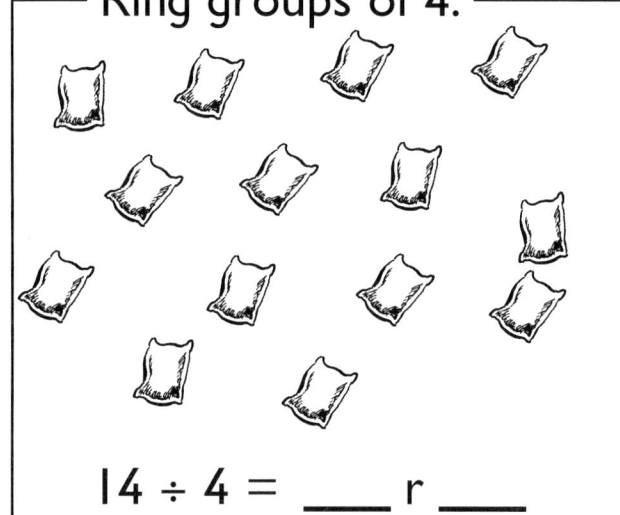

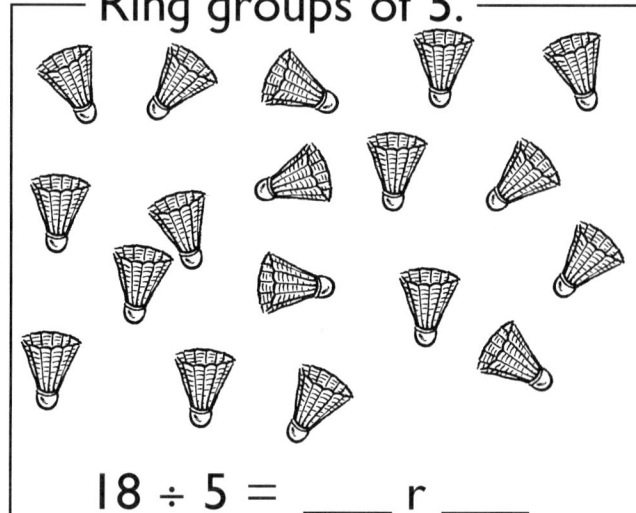

16 ÷ 3 = ___ r ___

Ring groups of 4.

14 ÷ 4 = ___ r ___

Ring groups of 5.

18 ÷ 5 = ___ r ___

36 ÷ 10 = ___ r ___   19 ÷ 2 = ___ r ___

Divide 29 by 3.   ___ r ___

17 divided by 4 =   ___ r ___

How many fives make 33?   ___ r ___

How many groups of 10 are in 85?   ___ r ___

Division • *Grouping with remainders*

Pupil Sheet 26

Name:

Use cubes.

Share equally.

15 ÷ 2 = ___ r ___

23 ÷ 3 = ___ r ___

23 ÷ ___ = ___ r ___

___ ÷ 5 = ___ r ___

___ ÷ ___ = ___ r ___

45 ÷ 10 = ___ r ___

25 ÷ 2 = ___ r ___

Divide 22 by 5. ___ r ___

7 divided by 2. = ___ r ___

Share equally 35 among 4. ___ r ___

Divide 20 equally among 3. ___ r ___

Division • *Sharing with remainders*

Name:

Match.

£4·32

£5·05

£3·51

£4·04

£4·43

£5·60

Money • *Counting coin collections with £2*

Name:

Pupil Sheet 28

Zoe: I have £1. I spend 45 pence.

Coins in Zoe's change.
_____

Amount left _____

Bob: I have £2. I spend 76 pence.

Coins in Bob's change.
_____

Amount left _____

Leela: I have £1. I spend 22 pence.

Coins in Leela's change.
_____

Amount left _____

Mel: I have £2. I spend 37 pence.

Coins in Mel's change.
_____

Amount left _____

Money • *Change from £1 and £2*

Name:

Pupil Sheet 29

Colour.

one tenth

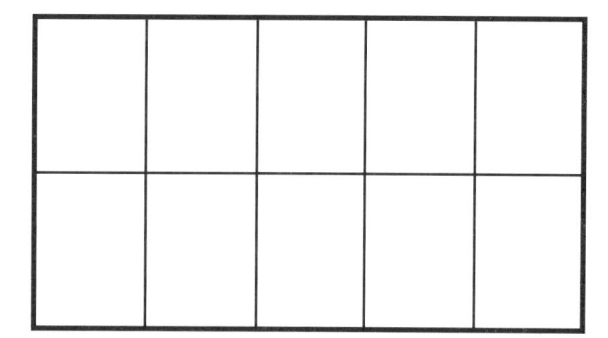

four tenths

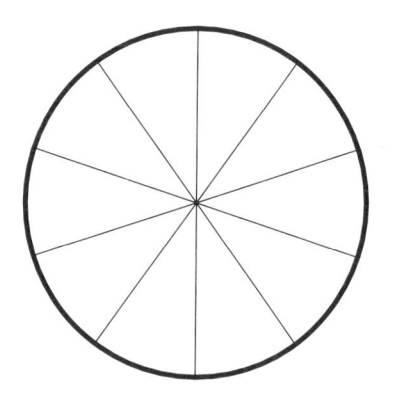

seven tenths

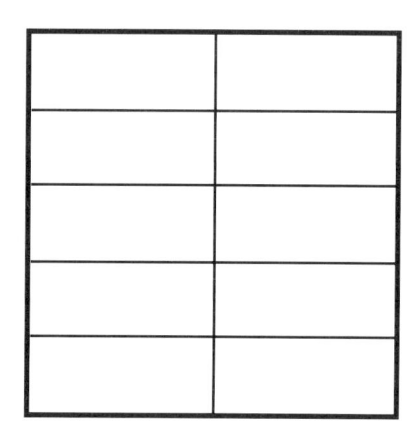

nine tenths

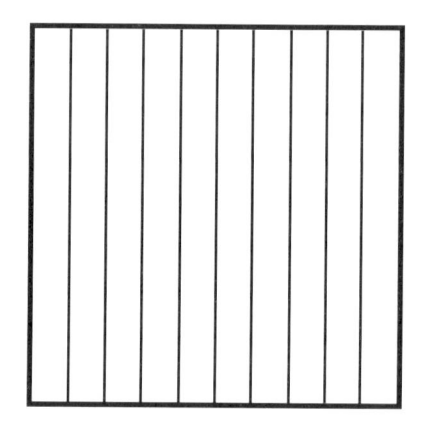

two tenths red
eight tenths blue

three tenths red
five tenths blue

Fractions • *Tenths*

Name:

Pupil Sheet 30

What fraction of each shape is missing?

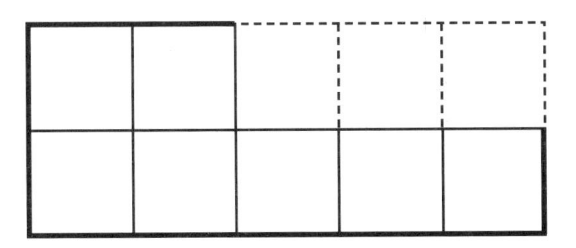

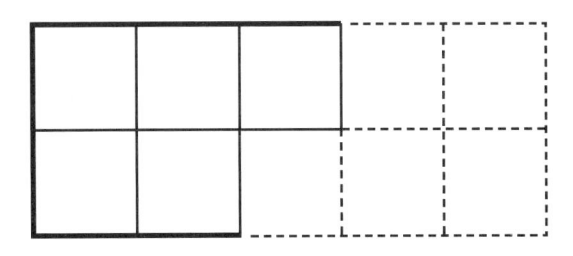

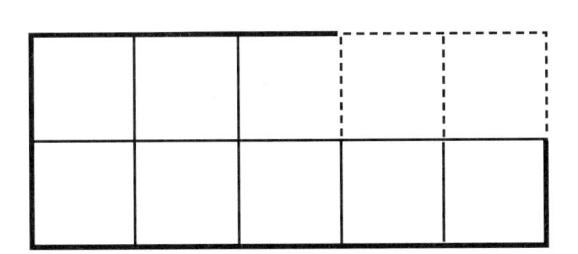

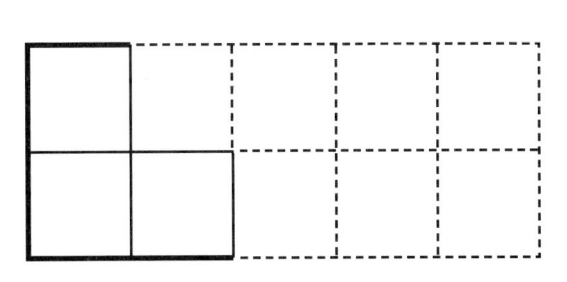

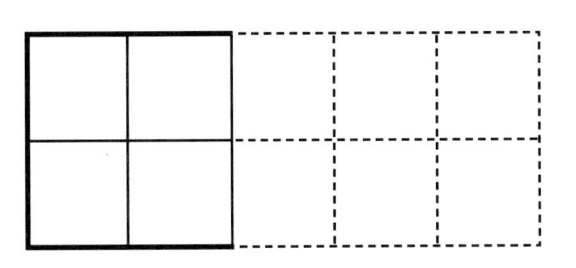

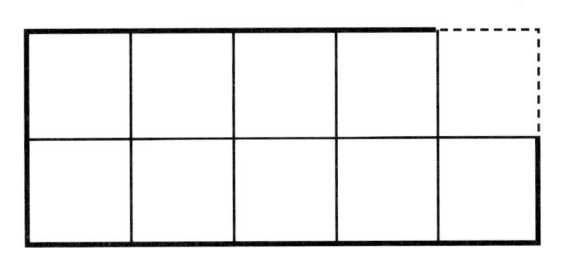

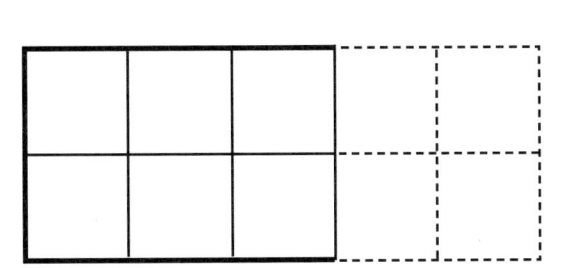

Fractions • *Tenths*

Name: _____

Match.

one half of 16   (6)   one quarter of 16

one tenth of 20   (4)   one half of 20

one half of 12   (8)   one tenth of 80

one quarter of 32   (2)   one quarter of 8

one tenth of 100   (10)   one half of 4

Complete.

one half of 10 = ___   one quarter of 36 = ___

one tenth of 90 = ___   one half of 14 = ___

one quarter of 28 = ___   one tenth of 10 = ___

$\frac{1}{2}$ of 8 = ___ $\frac{1}{2}$ of 18 = ___ $\frac{1}{2}$ of 2 = ___

$\frac{1}{4}$ of 20 = ___ $\frac{1}{4}$ of 40 = ___ $\frac{1}{4}$ of 12 = ___

$\frac{1}{10}$ of 50 = ___ $\frac{1}{10}$ of 60 = ___ $\frac{1}{10}$ of 30 = ___

Tick (✓) the larger. | $\frac{1}{4}$ of 24 | | $\frac{1}{10}$ of 70 |

Fractions • *Half of, quarter of, tenth of*

Pupil Sheet 31

Name: _____

Pupil Sheet 32

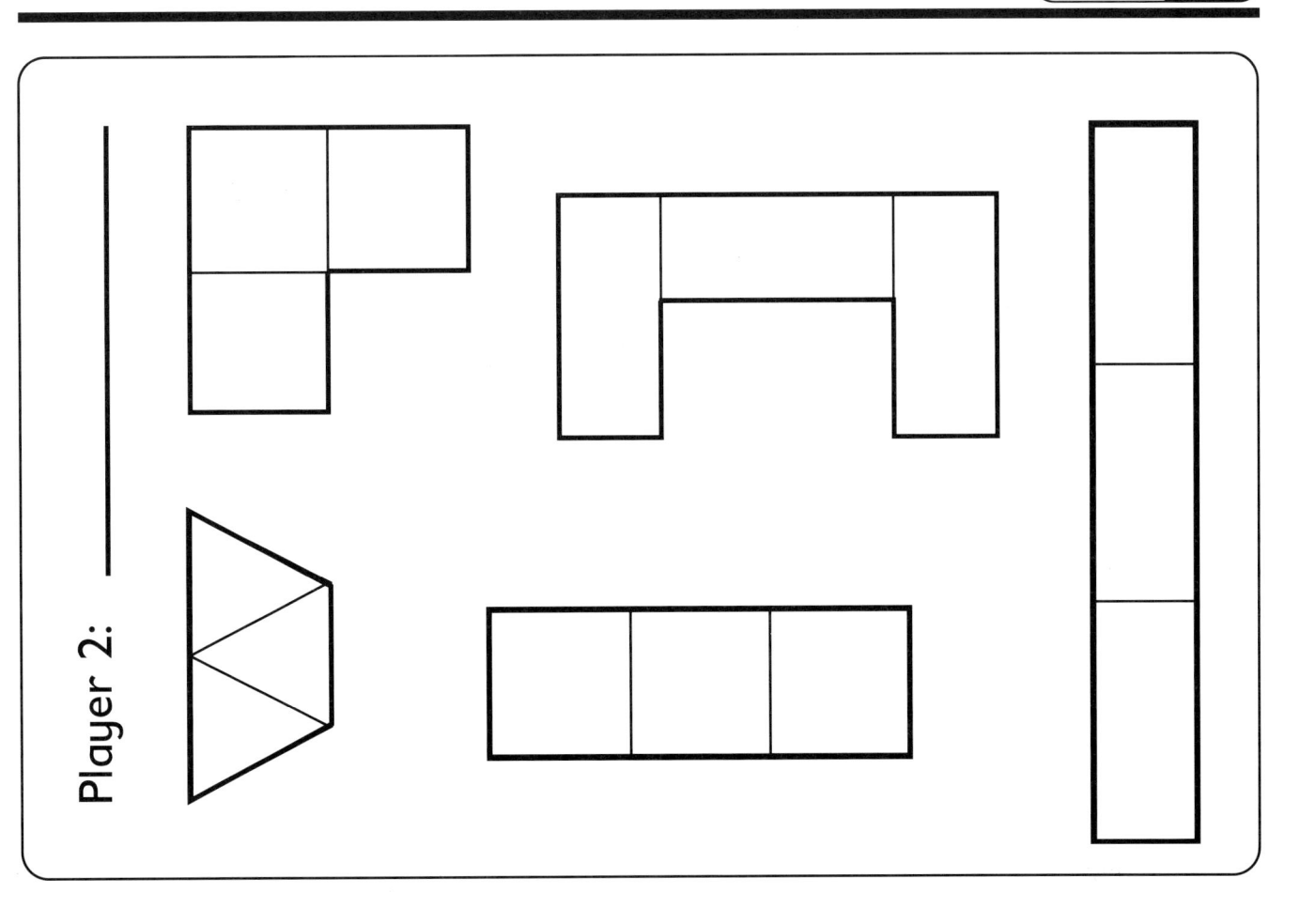

Player 2: _____

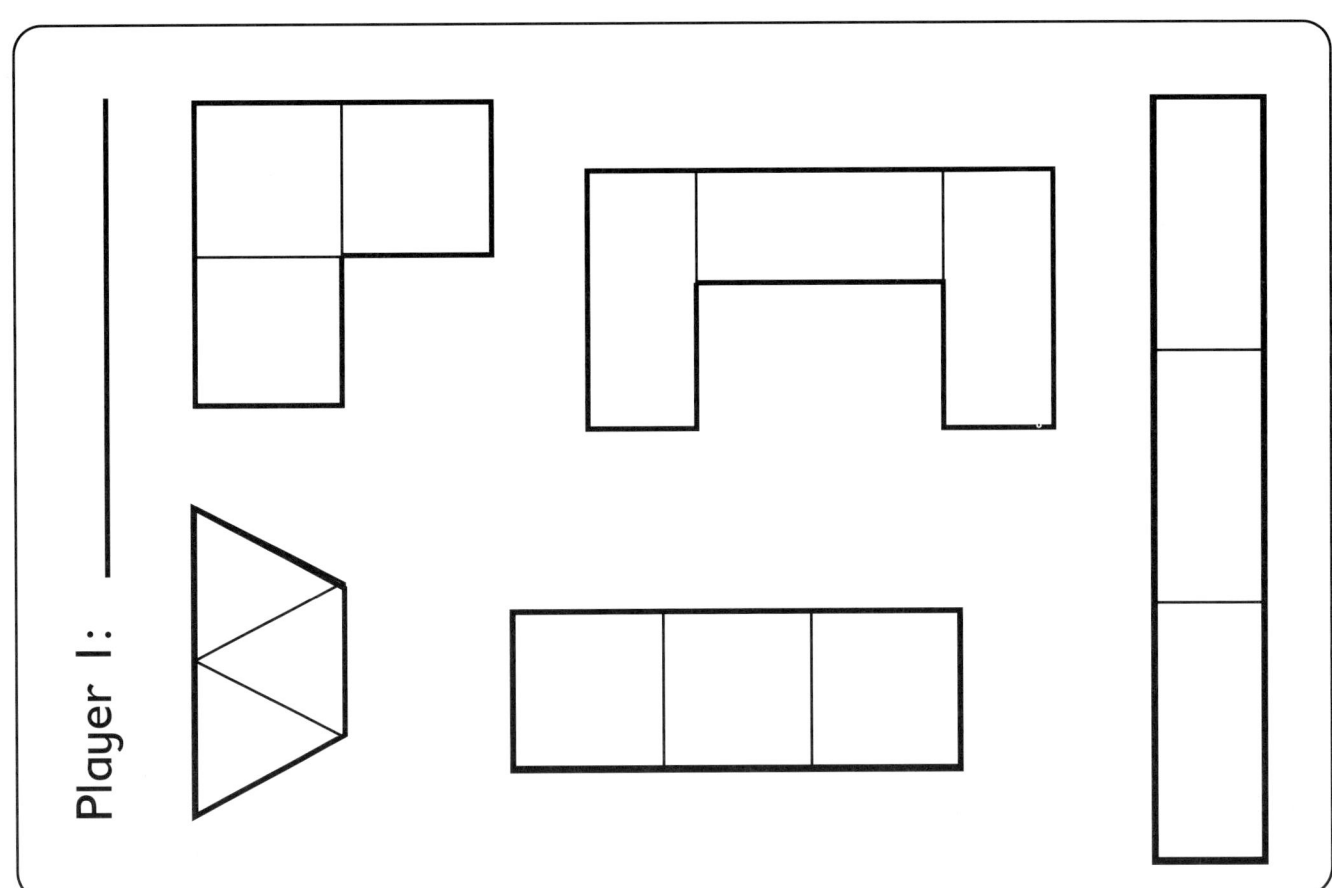

Player 1: _____

Fractions • *Thirds*

Name:

Colour.

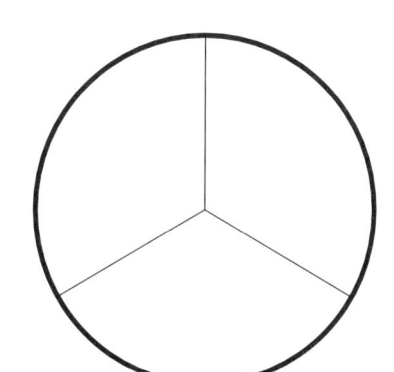

one third

one tenth

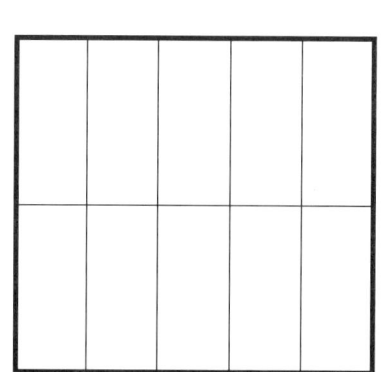

one fifth

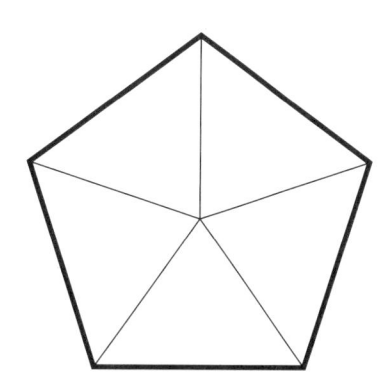

two thirds

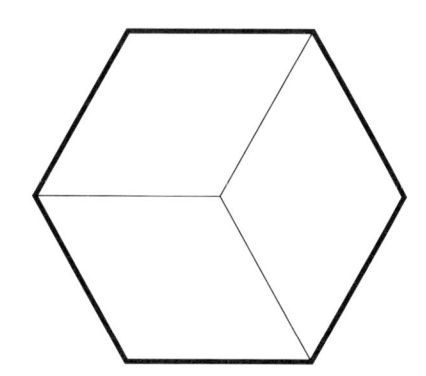

four fifths

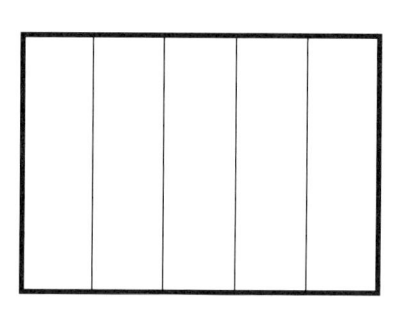

seven tenths

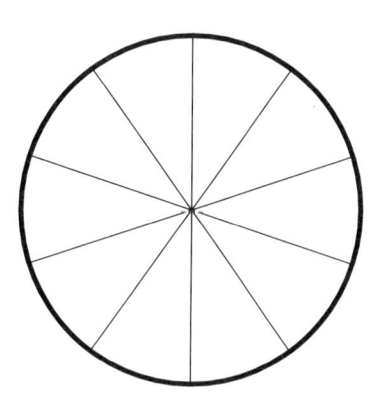

three fifths

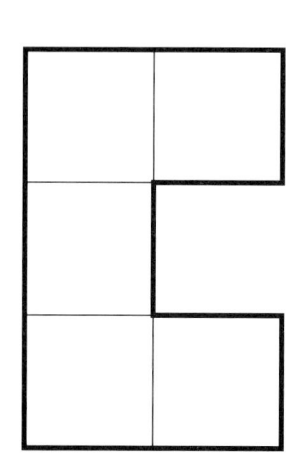

five tenths

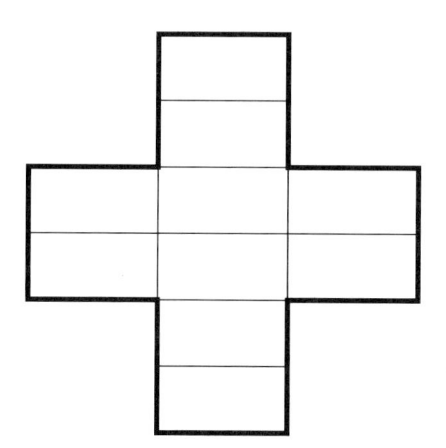

Fractions • *Tenths, thirds, fifths*

Name:

Pupil Sheet 34

What fraction of each shape is • shaded • not shaded?

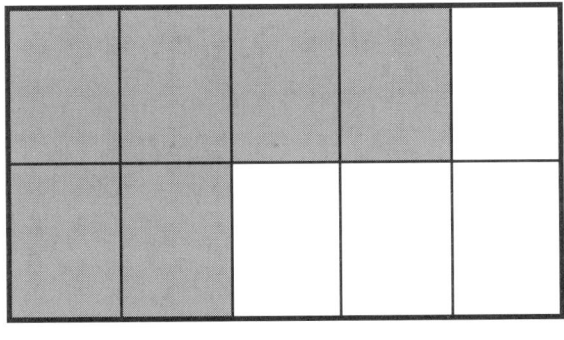

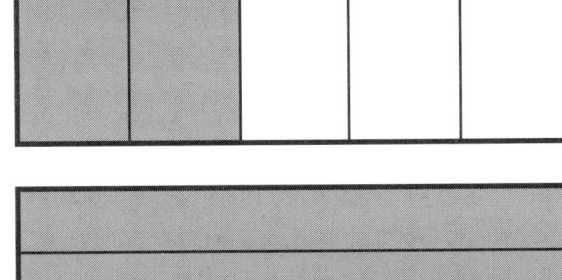

shaded _____

not shaded _____

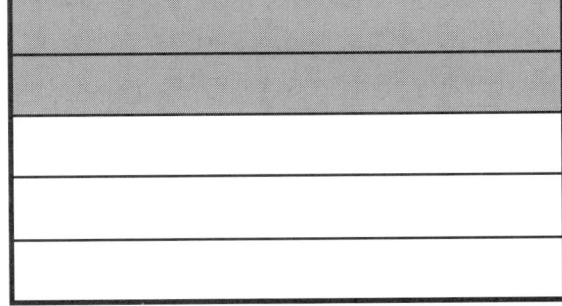

shaded _____

not shaded _____

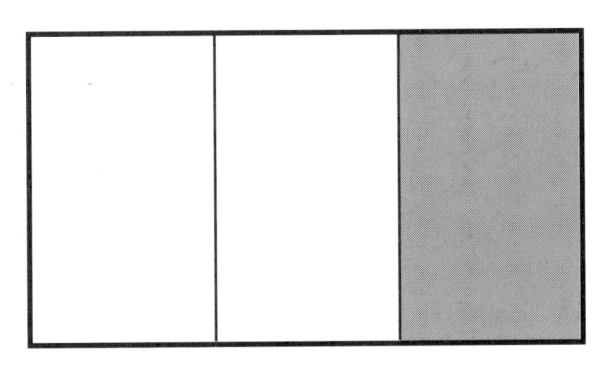

shaded _____

not shaded _____

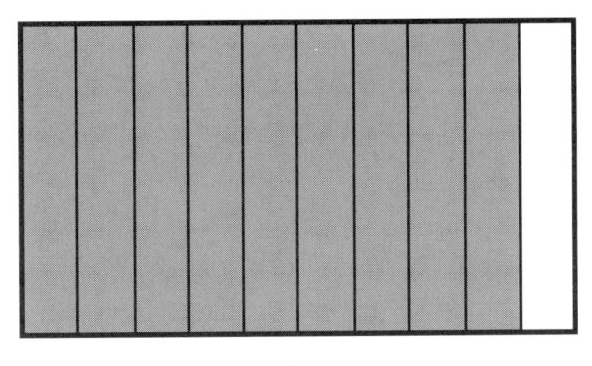

shaded _____

not shaded _____

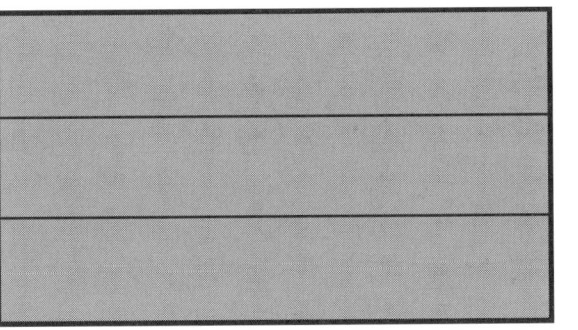

shaded _____

not shaded _____

Fractions • *Tenths, thirds, fifths*

Name: _____

Pupil Sheet 35

Find.

one fifth of 35 = ___   one third of 9 = ___

one fifth of 50 = ___   one third of 15 = ___

$\frac{1}{5}$ of 20 = ___ $\frac{1}{3}$ of 21 = ___ $\frac{1}{5}$ of 45 = ___

$\frac{1}{3}$ of 6 = ___ $\frac{1}{5}$ of 5 = ___ $\frac{1}{3}$ of 30 = ___

Match.

$\frac{1}{5}$ of 10   one third of 27

$\frac{1}{3}$ of 15   one fifth of 20

one fifth of 30 $\frac{1}{3}$ of 18

one third of 12 $\frac{1}{5}$ of 25

6, 4, 9, 2, 5

Tick (✓) the smaller. $\frac{1}{3}$ of 24 $\frac{1}{5}$ of 15

Fractions • Third, fifth of a set

New Heinemann Maths 3 © SPMG 2000

Name:

Pupil Sheet 36

5 + 6 = ☐
50 + 60 = ☐

8 + 4 = ☐
80 + 40 = ☐

7 + 9 = ☐
☐ + ☐ = ☐

9 + 9 = ☐
☐ + ☐ = ☐

6 + 8 = ☐
☐ + ☐ = ☐

4 + 9 = ☐
☐ + ☐ = ☐

8 + 9 = ☐
☐ + ☐ = ☐

7 + 8 = ☐
☐ + ☐ = ☐

Addition and subtraction to 1000 • *Adding multiples of 10*

Name:

Pupil Sheet 37

83 + 62 = ☐

92 + 28 = ☐

58 + 64 = ☐

84 + 47 = ☐

68 + 76 = ☐

75 + 37 = ☐

Addition and subtraction to 1000 • *Adding two-digit numbers, bridging 100*

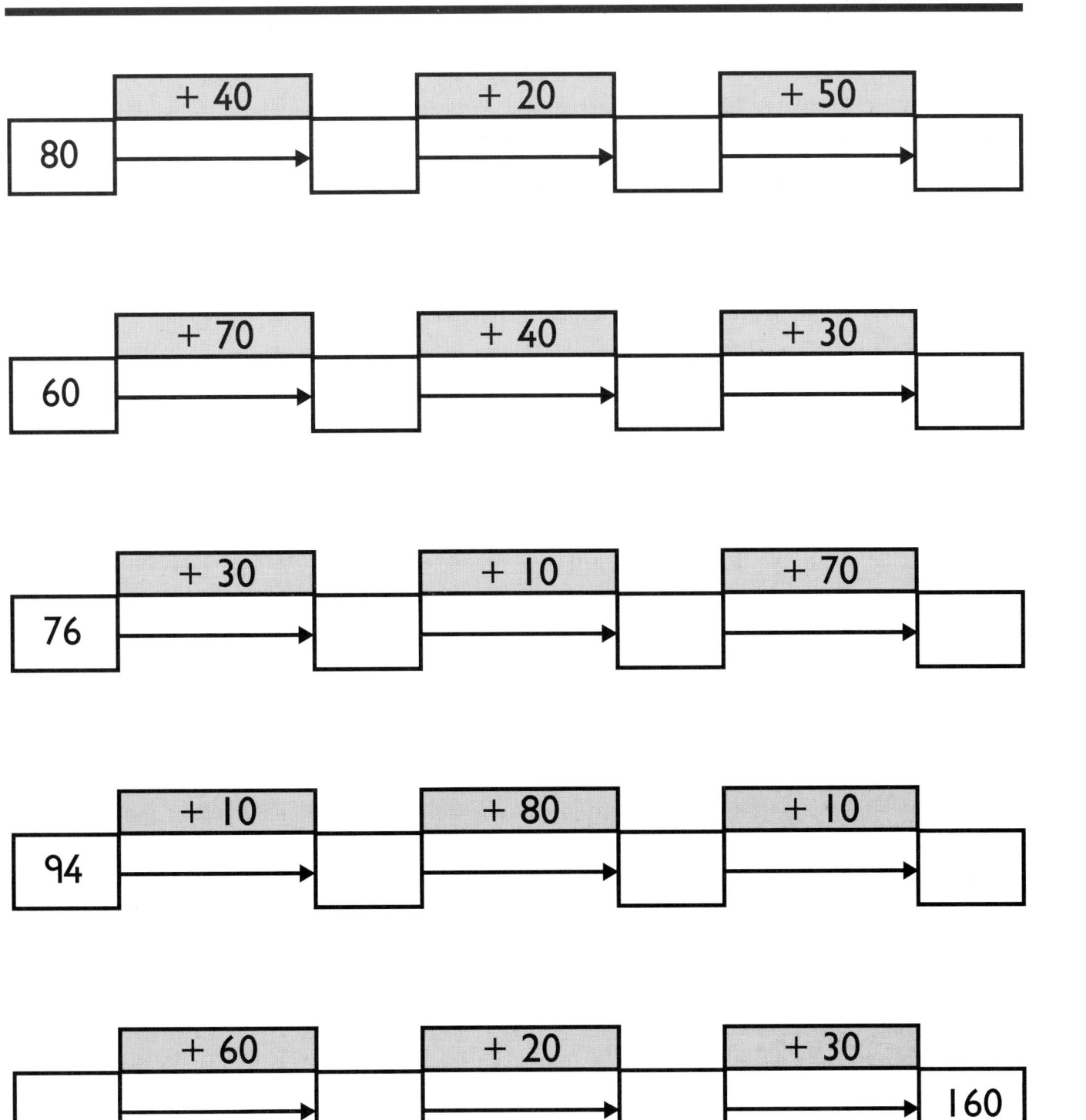

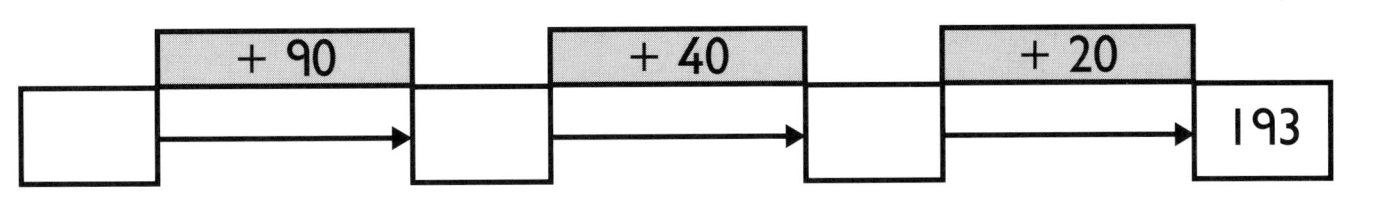

Pupil Sheet 38

Name:

80 → +40 → ☐ → +20 → ☐ → +50 → ☐

60 → +70 → ☐ → +40 → ☐ → +30 → ☐

76 → +30 → ☐ → +10 → ☐ → +70 → ☐

94 → +10 → ☐ → +80 → ☐ → +10 → ☐

☐ → +60 → ☐ → +20 → ☐ → +30 → 160

☐ → +90 → ☐ → +40 → ☐ → +20 → 193

Addition and subtraction to 1000 • Adding two-/three-digit numbers and multiples of 10

Pupil Sheet 39

Name:

13 − 6 = 7
130 − 60 = 70
135 − 60 = 75

12 − 7 =
120 − 70 =
128 − 70 =

11 − 4 =
110 − 40 =
112 − 40 =

10 − 2 =
100 − 20 =
103 − 20 =

14 − 9 =
140 − 90 =
144 − 90 =

15 − 6 =
150 − 60 =
159 − 60 =

16 − 8 =
160 − 80 =
167 − 80 =

15 − 5 =
150 − 50 =
157 − 50 =

Addition and subtraction to 1000 • *Subtracting a multiple of 10 from a three-digit number*

Name:

Pupil Sheet 40

Find the difference between

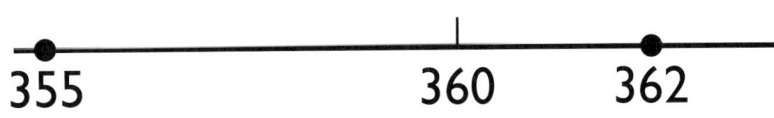

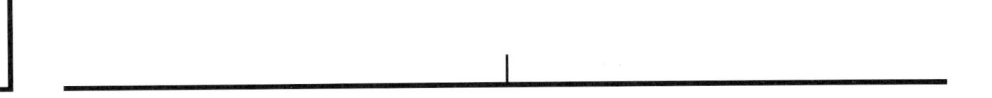

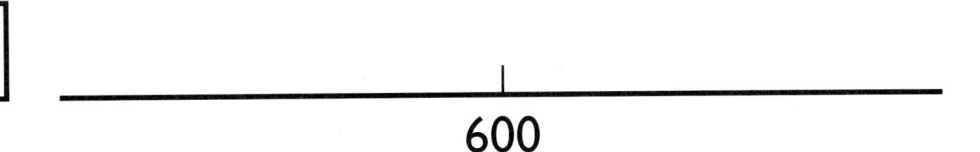

355 and 362 ☐   •————|————•————
           355   360   362

522 and 517 ☐   ————————|————————

601 and 598 ☐   ————————|————————
                         600

896 and 904 ☐   ————————|————————

| 202 | — a difference of ___ — | 211 |

| 300 | — a difference of ___ — | 293 |

| 696 | — a difference of ___ — | 705 |

| 1000 | — a difference of ___ — | 995 |

Addition and subtraction to 1000 • Finding small differences

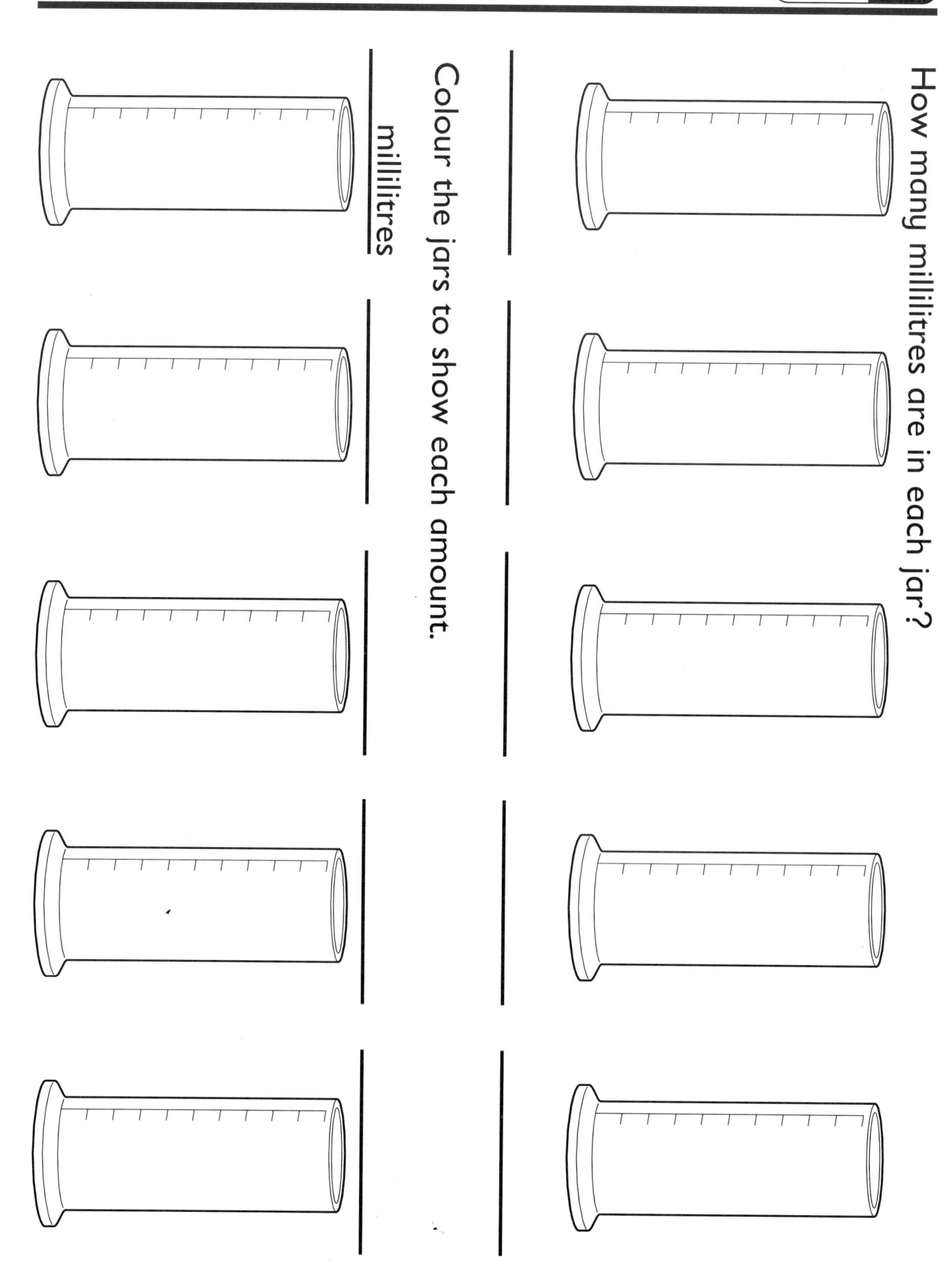

Name:

Pupil Sheet 41

How many millilitres are in each jar?

Colour the jars to show each amount.

_____ millilitres

Capacity • *Millilitres*

New Heinemann Maths 3 © SPMG 2000
Copying permitted for purchasing school only. This material is not copyright free.

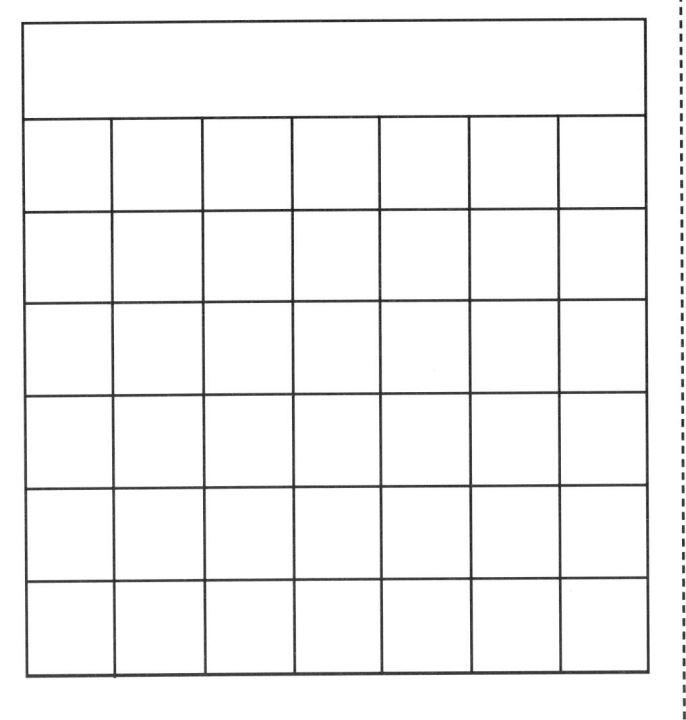

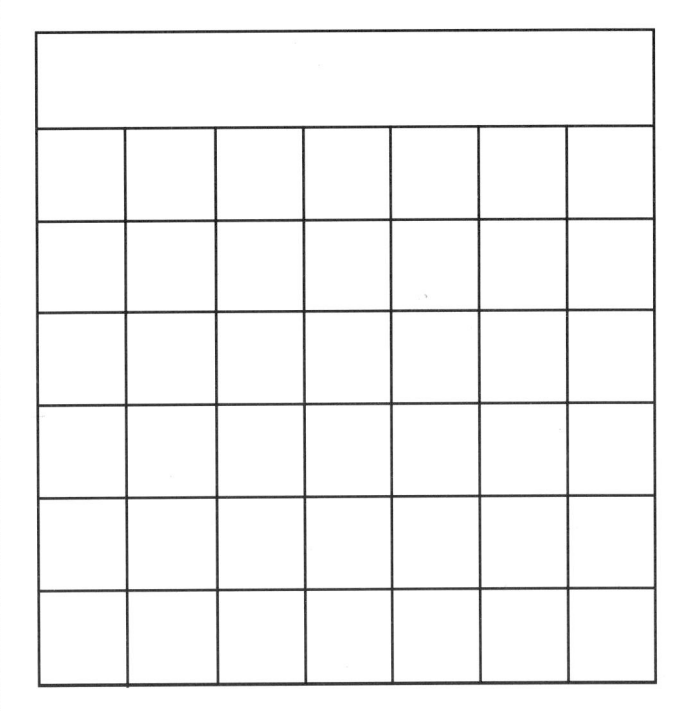

Name:

Pupil Sheet 42

January

Sun	Mon	Tue	Wed	Thu	Fri	Sat
	1	2	3	4	5	6
7	8	9	10	11	12	13
14	15	16	17	18	19	20
21	22	23	24	25	26	27
28	29	30	31			

January

Sun	Mon	Tue	Wed	Thu	Fri	Sat
	1	2	3	4	5	6
7	8	9	10	11	12	13
14	15	16	17	18	19	20
21	22	23	24	25	26	27
28	29	30	31			

Time • *The calendar*

Name:

Pupil Sheet 43

Time • *Writing times*

Name:

Pupil Sheet 44

Time • *Minutes past and to (digital)*

New Heinemann Maths 3 © SPMG 2000
Copying permitted for purchasing school only. This material is not copyright free.

Name:

Pupil Sheet 45

Colour the time

Clock	5 minutes before	two clocks
Clock	20 minutes after	two clocks
Clock	30 minutes before	two clocks
Clock	3 hours after	two clocks
 10:10	20 minutes before	 10:00 / 9:50
 6:40	25 minutes after	 7:05 / 6:55

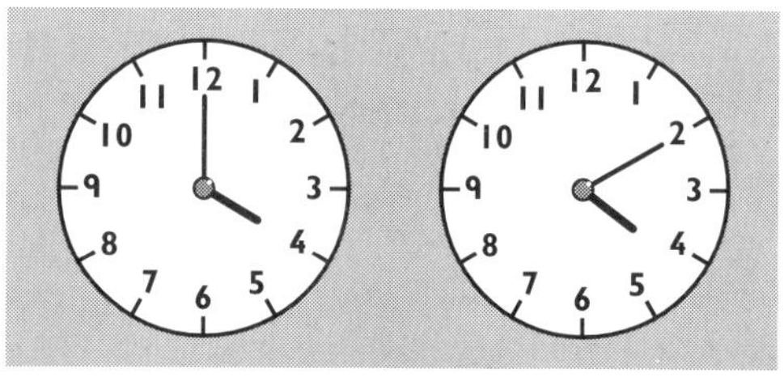

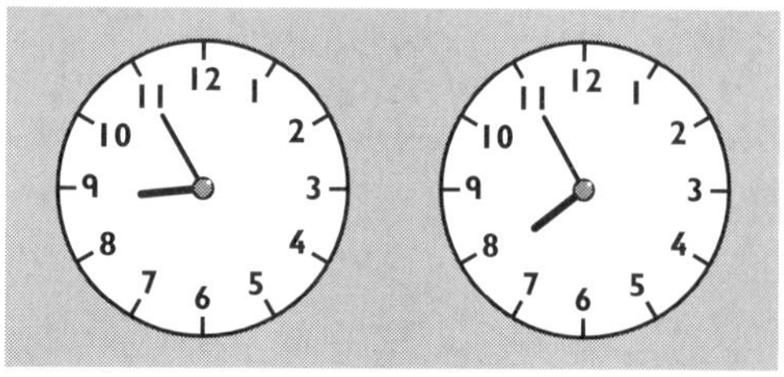

Time • *Minutes/hours before/after*

Name:

Pupil Sheet 46

Use two △ △
Make a hexagon.

Use two ▭ ▭
Make a quadrilateral.

Use two ▭ ▭
Make a hexagon.

Use two ▭ ▭
Make an octagon.

2D shape • *Quadrilaterals, hexagons, octagons*

Name:

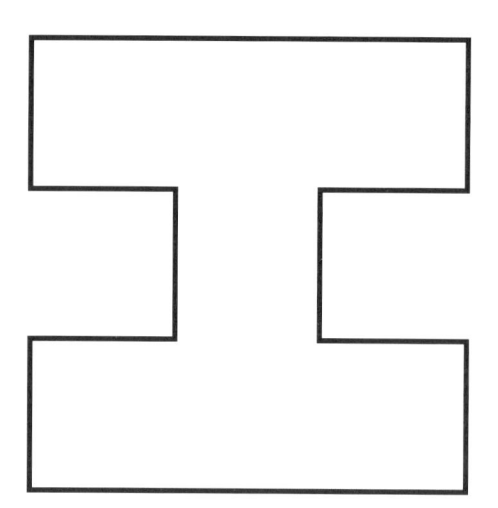

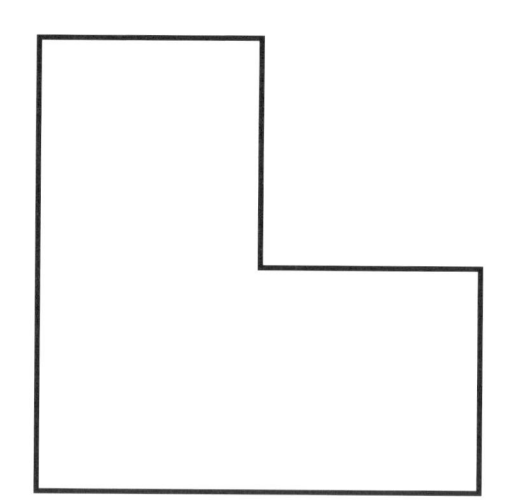

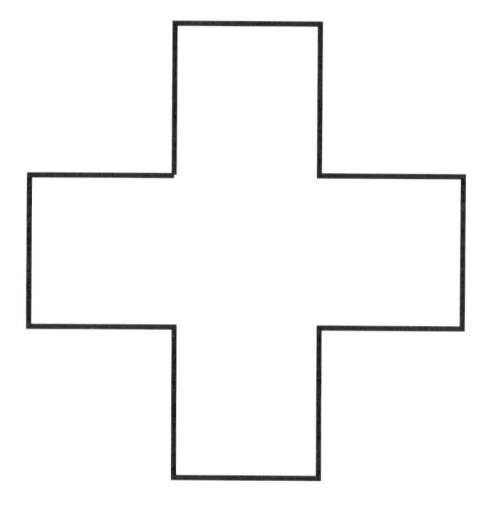

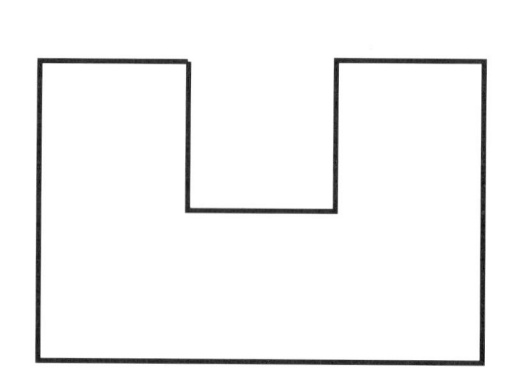

Pupil Sheet 47

2D shape • *Strip patterns*

New Heinemann Maths 3 © SPMG 2000
Copying permitted for purchasing school only. This material is not copyright free.

Name:

Pupil Sheet 48

- ✂ - -

2D shape • *Using and applying*

New Heinemann Maths 3 © SPMG 2000
Copying permitted for purchasing school only. This material is not copyright free.

Name:

Pupil Sheet 49

	A	B	C	D	E	F	G
5							
4							
3							
2							
1							

2D shape • *Grid references*

New Heinemann Maths 3 © *SPMG 2000*
Copying permitted for purchasing school only. This material is not copyright free.

Name:

Colour
- right angles red
- angles **larger** than a right angle blue
- angles **smaller** than a right angle green.

Pupil Sheet 50

2D shape • *Angles*

New Heinemann Maths 3 © SPMG 2000
Copying permitted for purchasing school only. This material is not copyright free.

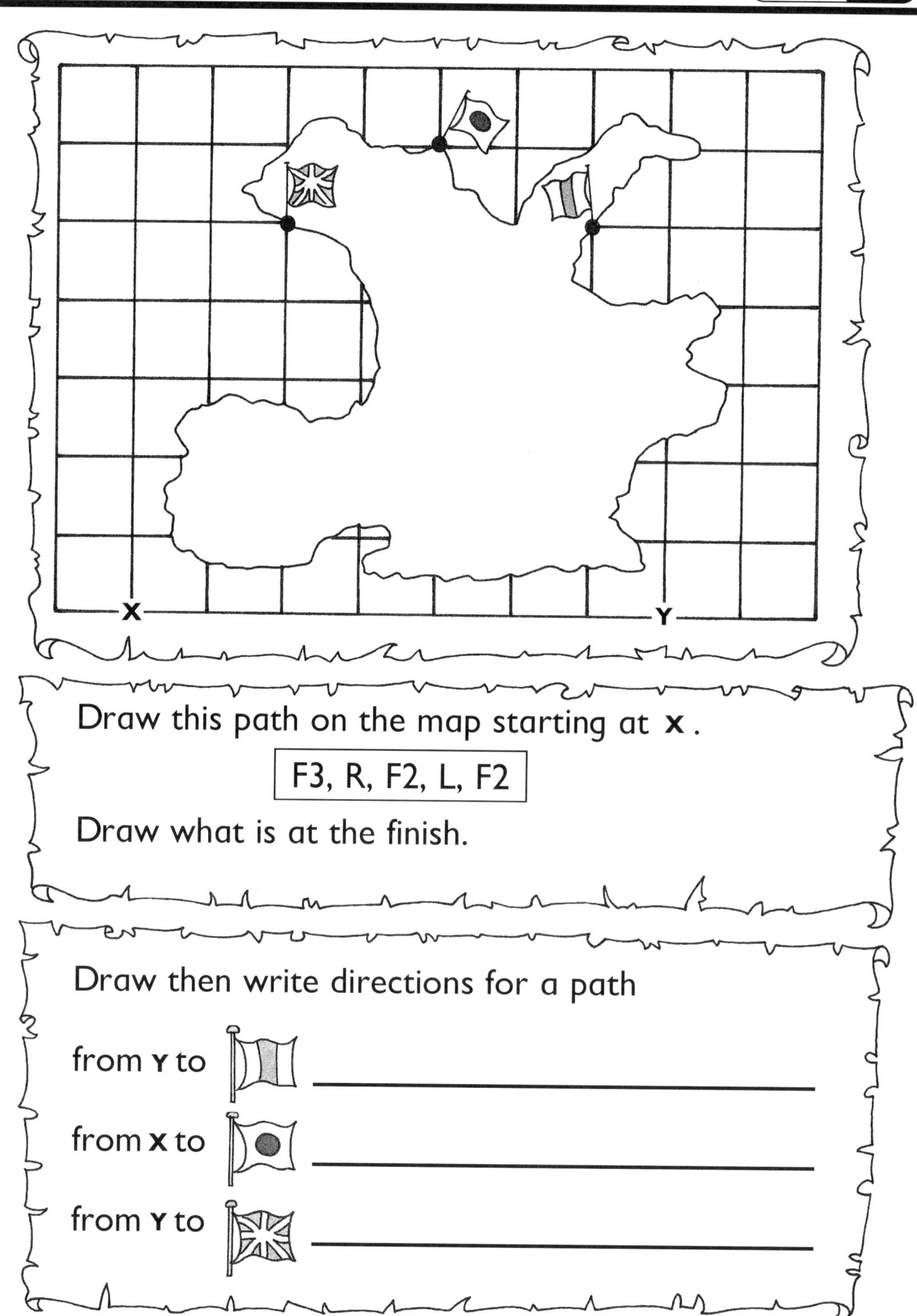

Name: _____ Pupil Sheet **51**

Draw this path on the map starting at **x**.

F3, R, F2, L, F2

Draw what is at the finish.

Draw then write directions for a path

from **Y** to 🚩 _____

from **x** to 🚩 _____

from **Y** to 🚩 _____

2D shape • *Position and movement*

Name:

Pupil Sheet 52

Cut out these shapes.
Use 2 shapes each time to make new symmetrical shapes.

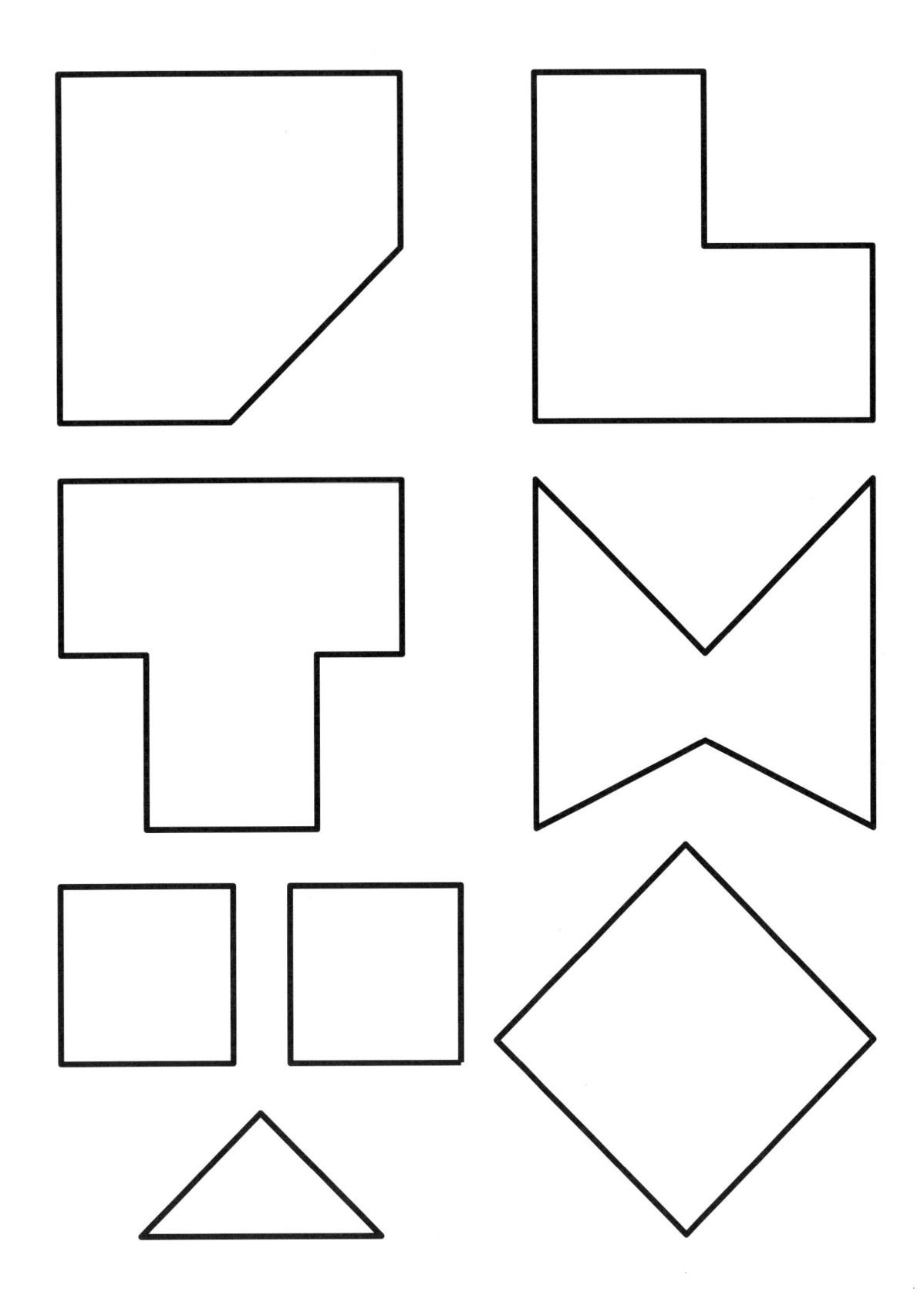

2D shape • *Symmetry*

Name:

Pupil Sheet 53

Draw and colour a in each box.

Data handling • *Carroll diagram*

New Heinemann Maths 3 © SPMG 2000
Copying permitted for purchasing school only. This material is not copyright free.

Name:

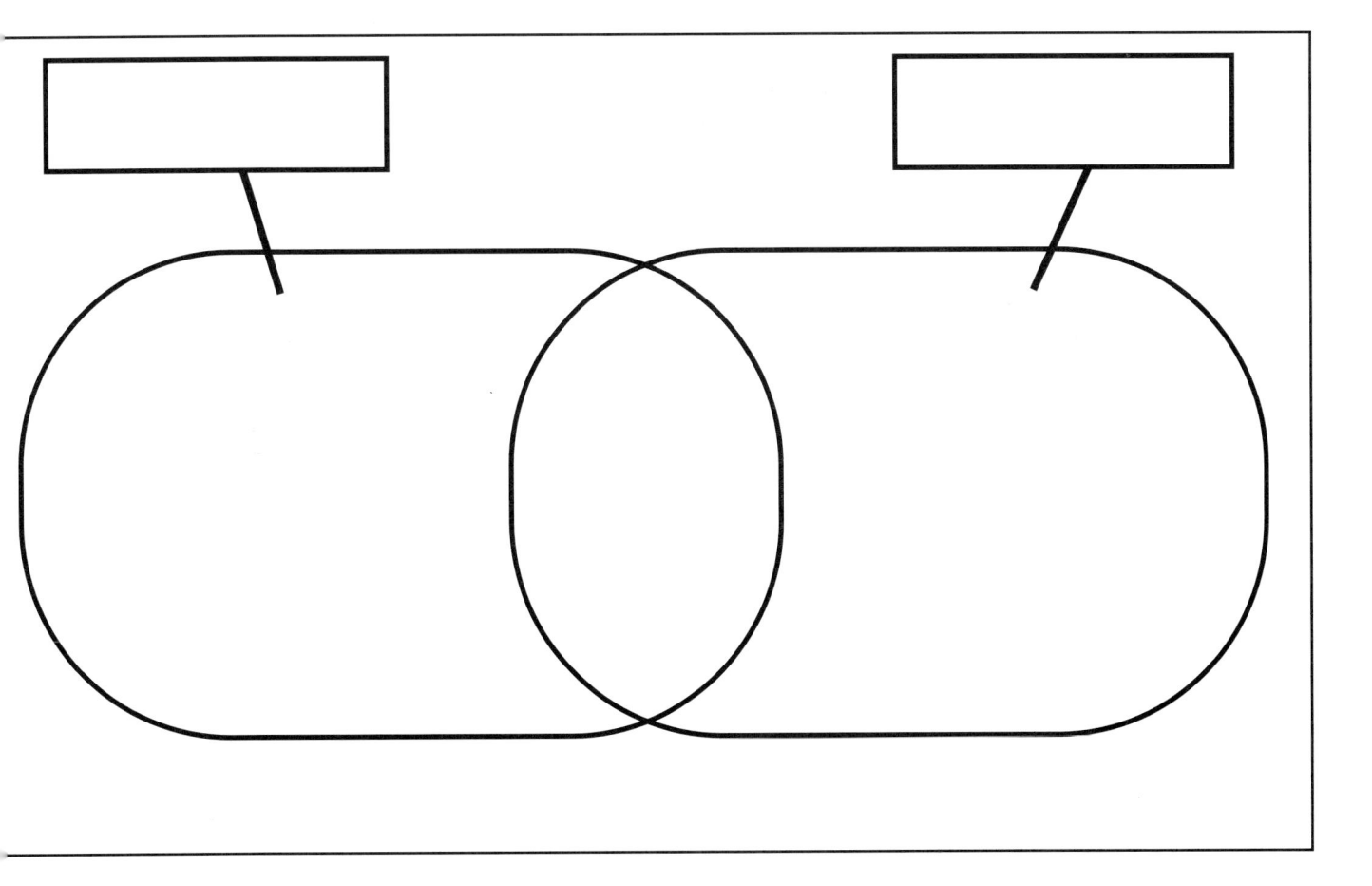

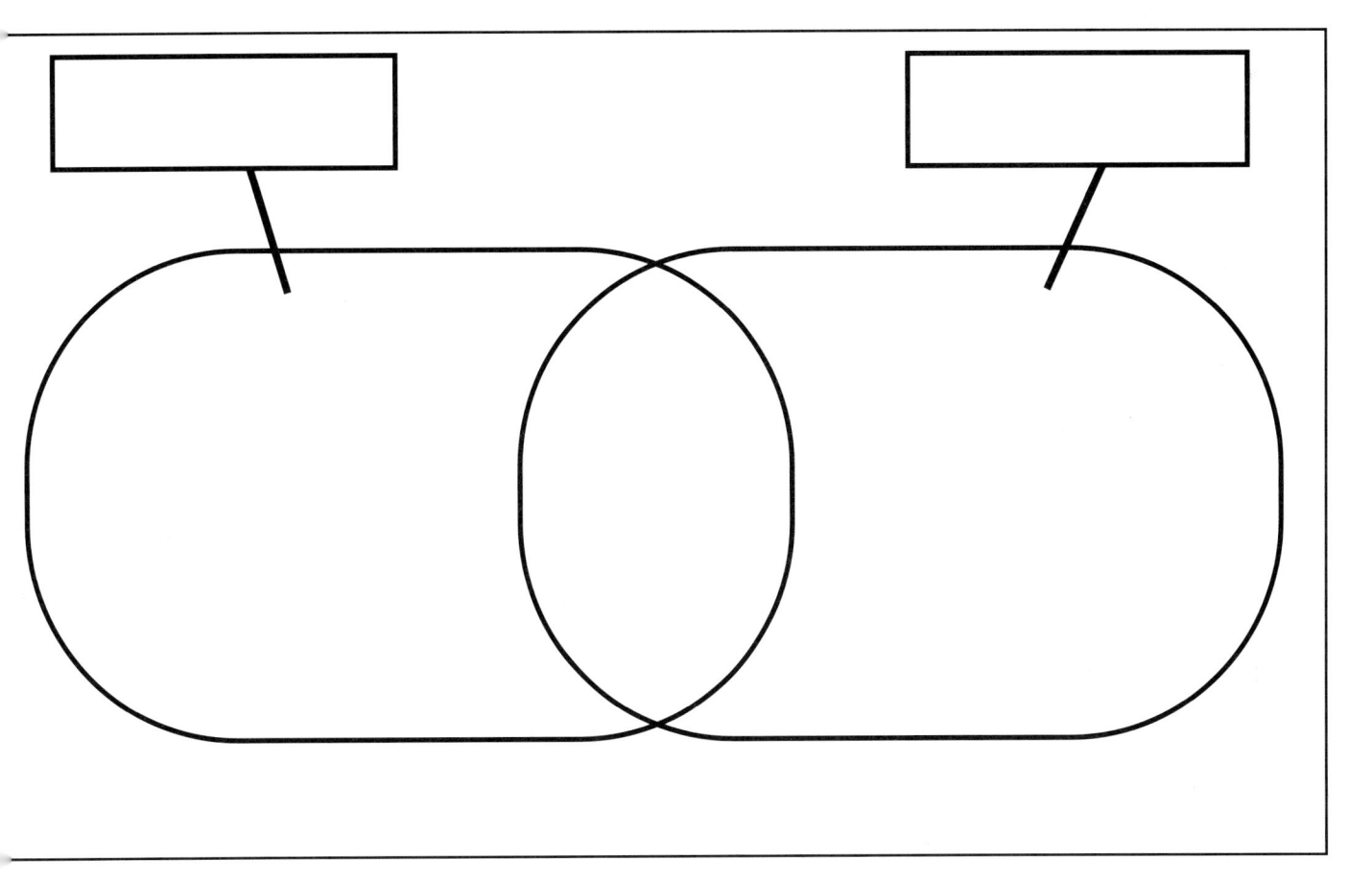

Pupil Sheet 54

Data handling • *Venn diagrams*

Home Activities

Numbers to 1000, then 10000

Home Activity **1**	Sequences to 1000
Home Activity **2**	Adding/Subtracting 10s, 100s
Home Activity **3**	Ordering numbers

Addition and Subtraction to 100

Home Activity **4**	Addition facts to 20
Home Activity **5**	Adding a single-digit number
Home Activity **6**	Adding two-digit numbers
Home Activity **7**	Mental addition
Home Activity **8**	Subtraction facts to 20
Home Activity **9**	Subtracting two-digit numbers
Home Activity **10**	Subtracting a single digit
Home Activity **11**	Subtracting a teens number

Multiplication and Division

Home Activity **12**	Two, five and ten times tables
Home Activity **13**	Three times table
Home Activity **14**	Four times table
Home Activity **15**	Multiplying by 2, 3, 4, 5 and 10
Home Activity **16**	Multiplying 2-digit numbers
Home Activity **17**	Dividing by 2 and 10
Home Activity **18**	Dividing by 5
Home Activity **19**	Dividing by 3
Home Activity **20**	Dividing by 4
Home Activity **21**	Dividing by 2, 3, 4, 5 and 10

Money

Home Activity **22**	Money
Home Activity **23**	£5, £10 and £20 notes

Fractions

Home Activity **24**	Tenths, '1/10 of'
Home Activity **25**	Thirds, fifths

Time

Home Activity **26**	Writing times
Home Activity **27**	Time intervals
Home Activity **28**	Domino cards for Home Activity 2
Home Activity **29**	Number cards for Home Activities 3 and 8
Home Activity **30**	Number cards for Home Activity 8
Home Activity **31**	Division cards for Home Activity 21

Home Activities • *Contents*

Name:

Home Activity 1

Sequences to 1000

Hats: 500, 640, 700, 609, 649

What is my hat number?

My number is
- between 600 and 700
- less than 640.

My number is _____ .

My number is
- between 630 and 660
- greater than 640.

My number is _____ .

Choose one of the other hat numbers.
Write a clue for it.

My number is

My number is _____ .

Write the number

after 283 _____   before 400 _____

one more than 693 _____   one less than 900. _____

Write **two** numbers between 810 and 825. _____ _____

Name:

Use Home Sheet 28 | **Home Activity 2**

Adding/Subtracting 10s, 100s

- Shuffle the domino cards and place them face up on the table.

Choose a card and place it in front of you.

The child chooses, for example: | 805 | 10 more than 250 |

What is the number ten more than two hundred and fifty?

Two hundred and sixty.

Find the card with two hundred and sixty. Put it beside the other card.

The child places | 260 | 100 less than 806 |

- Repeat until all eight cards have been placed in a line.

---

186 + 10 = ☐   342 + 30 = ☐   726 + 50 = ☐

962 − 10 = ☐   857 − 20 = ☐   590 − 60 = ☐

216 add 100 ☐   800 subtract 100 ☐

382 add 300 ☐   650 minus 400 ☐

Help — The child can count on or count back in 10s or 100s, using fingers, if necessary, to keep track.

Numbers to 1000: Activity Book page 10

Ordering numbers

Use Home Sheet 29 — **Home Activity 3**

- Scatter the number cards on the table.

Choose three cards.
Make me a number.
Tell me the number.
Five hundred and thirty-eight.

Ask the child to try to make five other numbers with the **same** three cards.

- Repeat for different sets of three cards.

Help Using 3, 8 and 5 the six possible numbers are: 385 358 538 583 835 853. Ask the child to write the numbers each time to help keep track.

---

471   322   654   368   259   437

Write

the largest number _____   the smallest number _____

the number between   the number between
368 and 471 _____   368 and 259. _____

Write the numbers in order. Start with the largest.

____ ____ ____ ____ ____ ____

Numbers to 1000: Activity Book page 17

Name:

Addition facts to 20

- Ask the child these different types of addition questions:

  Add nine and five.   Eight add six?
  What is four and eight?   Five plus seven?

  Repeat for these additions:

  3 + 9   4 + 10   8 + 5   7 + 7
  9 + 8   7 + 6   10 + 8   2 + 9

- Ask the child to point to and say pairs of numbers which add to make 15.

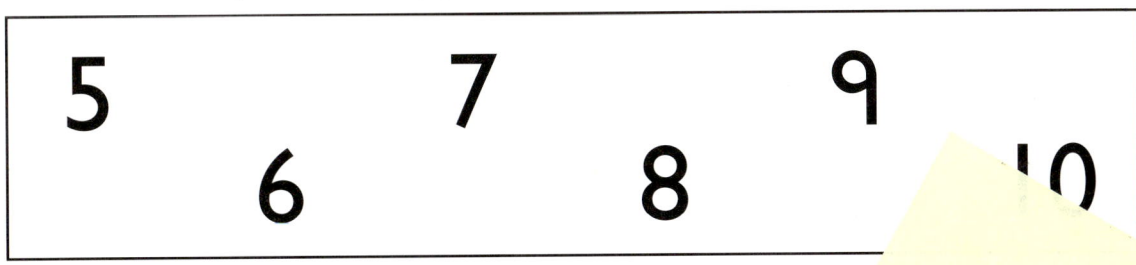

  5   7   9
     6   8   10

8 + 7 = ☐   5 + 6 = ☐

5 + 10 = ☐   6 + 9 = ☐   7 + 4 = ☐

Make 13.
9 +
+ 6

Make 17.
10 +
+ 8

Addition to 100: Textbook page 2

Home Activity 4

Home Activity 5

Adding a single-digit number

"Thirteen add four."

"Seventeen."

Ask these questions to practise addition facts to 20.

- Twelve add six?
- Add two to fifteen.
- Two more than sixteen?
- One more than eighteen?
- Five plus eleven?
- Add three and fourteen.

---

5 + 14 = ☐   8 + 11 = ☐   16 + 3 = ☐

17 + 2 = ☐   15 + 4 = ☐   6 + 12 = ☐

Make 20.

14 + ☐   ☐ + 17   ☐ + 8

43 + 5 = ☐   4 + 64 = ☐   52 + 7 = ☐

31 + 6 = ☐   7 + 23 = ☐   85 + 4 = ☐

Addition to 100: Textbook page 5

Name:

Adding two-digit numbers

Home Activity 6

The child and adult, in turn:
– choose a box and **mentally** add the numbers
– if correct, colour the box (each using a different colour).

35 + 14	23 + 41	34 + 22	51 + 28	16 + 32
55 + 45	50 + 37	15 + 85	54 + 31	27 + 62
83 + 16	35 + 65	60 + 29	17 + 51	43 + 53

37 + 9 = ☐   28 + 19 = ☐   44 + 39 = ☐

56 + 11 = ☐   33 + 21 = ☐   27 + 51 = ☐

29 + 35 = ☐   44 + 31 = ☐   19 + 21 = ☐

Help

To add 31, I add 30 then add 1.
To add 29, I add 30 then subtract 1.

Addition to 100: Textbook page 10

Name:

Home Activity 7

Mental addition

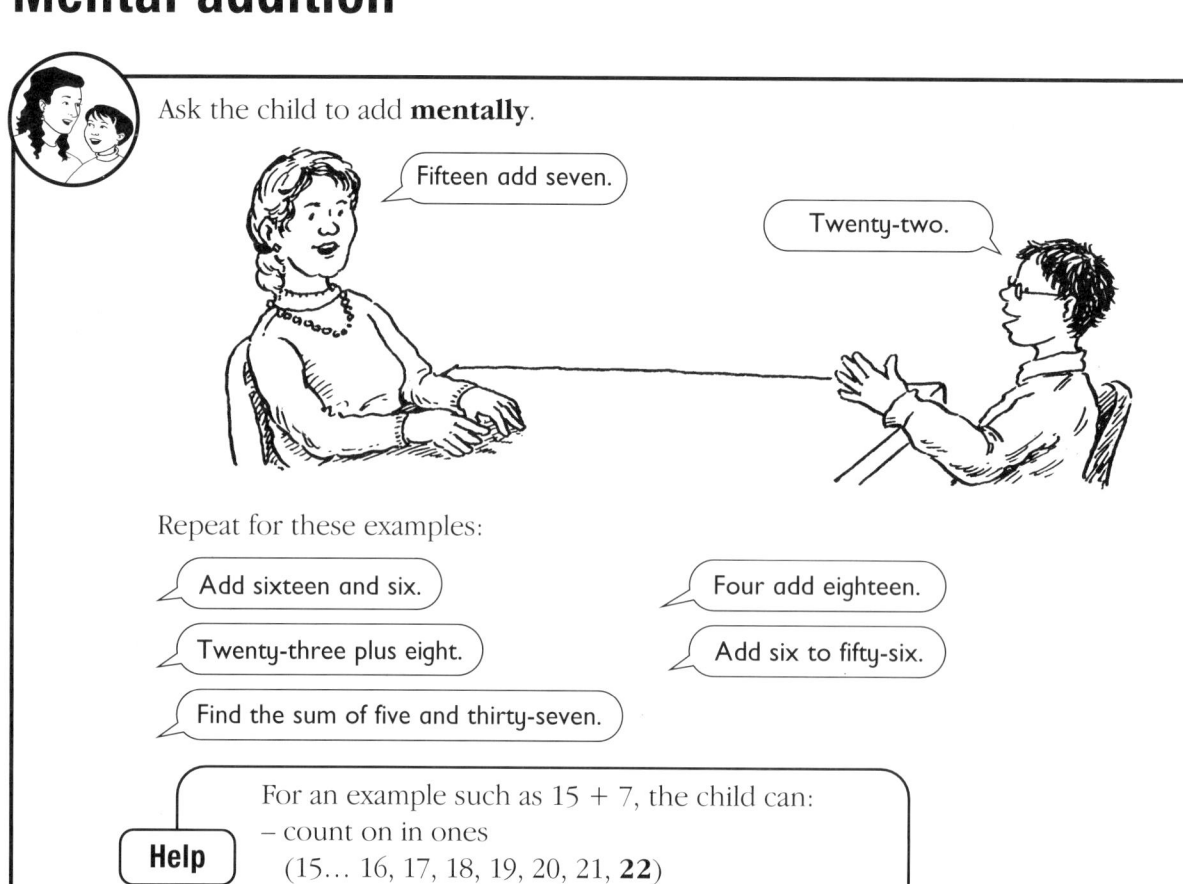

Ask the child to add **mentally**.

"Fifteen add seven."

"Twenty-two."

Repeat for these examples:

- Add sixteen and six.
- Twenty-three plus eight.
- Find the sum of five and thirty-seven.
- Four add eighteen.
- Add six to fifty-six.

Help

For an example such as 15 + 7, the child can:
– count on in ones
  (15... 16, 17, 18, 19, 20, 21, **22**)
– add 5 to make 20, then add 2 to make **22**.

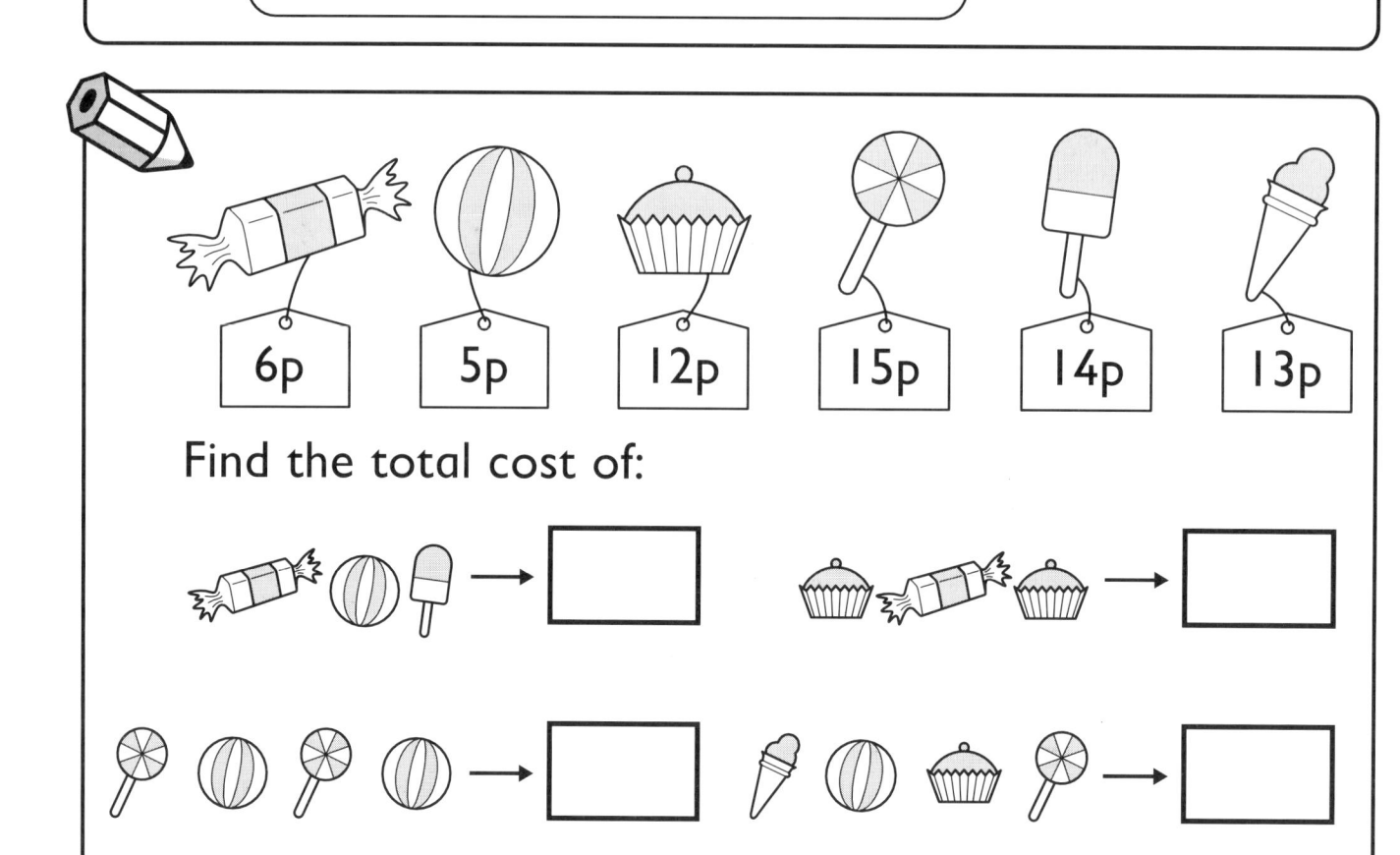

6p   5p   12p   15p   14p   13p

Find the total cost of:

Addition to 100: Textbook page 14

New Heinemann Maths 3 © SPMG 2000
Copying permitted for purchasing school only. This material is not copyright free.

Name:

Use Home Sheets **29 & 30** Home Activity **8**

Subtraction facts to 20

- Place the numbers cards face down on the table.

"Choose two cards."

"What is the difference between the numbers?"

"Thirteen."

(Once used, cards are replaced face up.)

- Repeat for other pairs of cards.

Subtract 10 from 14. ____

Take 17 from 19. ____

13 fewer than 16 ____

14 less than 20 ____

15 minus 9 ____

18 subtract 11 ____

17 take away 12 ____

20 minus 15 ____

Find the difference in price between

____ and ____ £ ____   £12

____ and ____ £ ____   £20

____ and ____ £ ____   £16

Subtraction to 100: Textbook page 20

Home Activity 9

Name:

Subtracting two-digit numbers

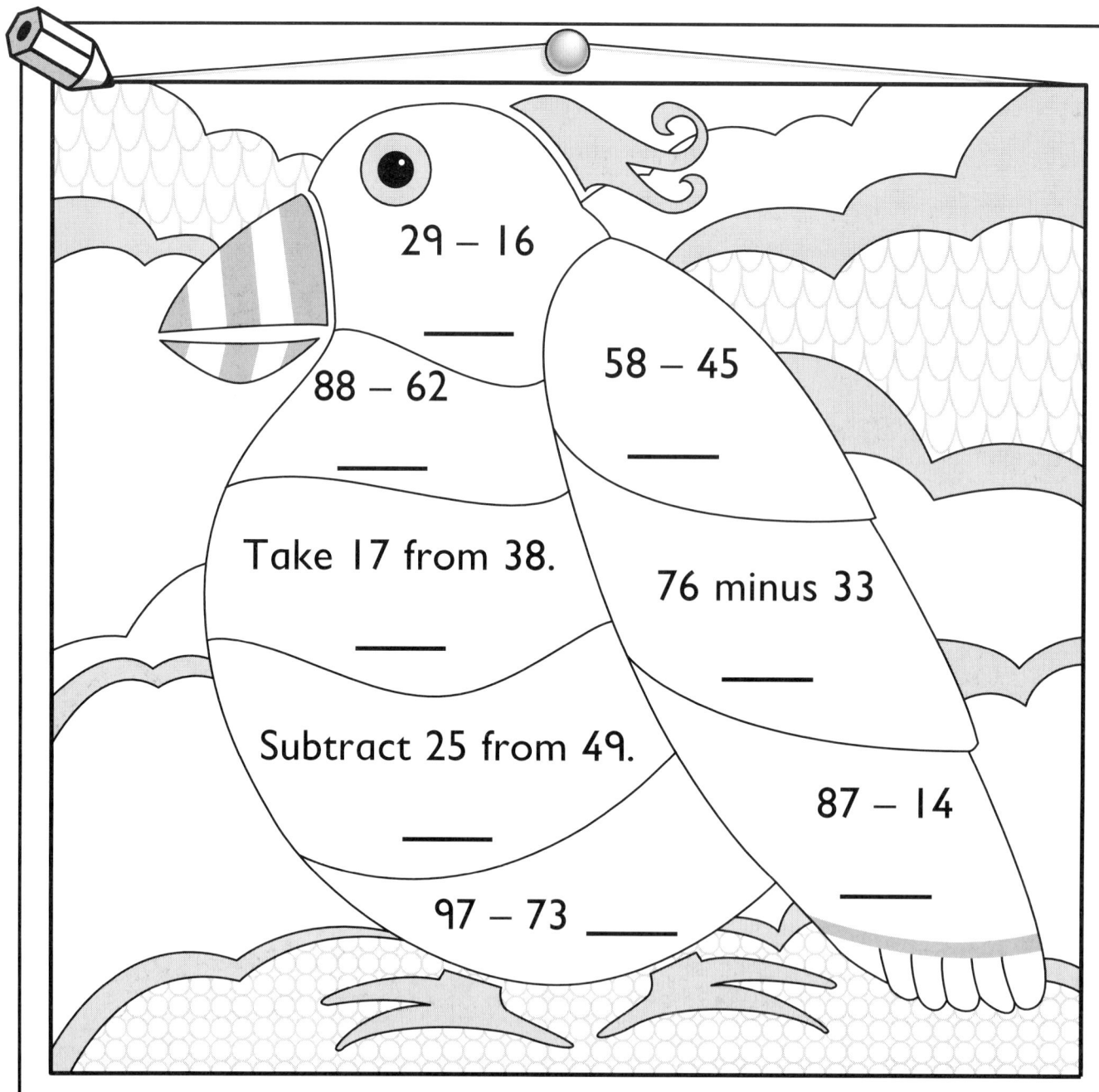

29 – 16 ___

88 – 62 ___

58 – 45 ___

Take 17 from 38. ___

76 minus 33 ___

Subtract 25 from 49. ___

87 – 14 ___

97 – 73 ___

Colour the parts which have answers with the
- units digit three ⟶ blue
- tens digit two ⟶ red.

Colour the other parts green.

Help Remind the child to subtract mentally in two steps. For 87 – 34:
step one 87 take away 30 ⟶ 57   *step two* 57 take away 4 ⟶ 53

Subtraction to 100: Textbook page 25

Subtracting a single digit

Home Activity 10

Use a die and, for each player, a counter. Start at the snake's head.
In turn, each player:
- rolls the die to find how many boxes along the track to move their finger
- finds mentally the answer to the subtraction in that box
- if correct, moves their counter to the box
- if wrong, does not move.

The first player to reach Tail wins.

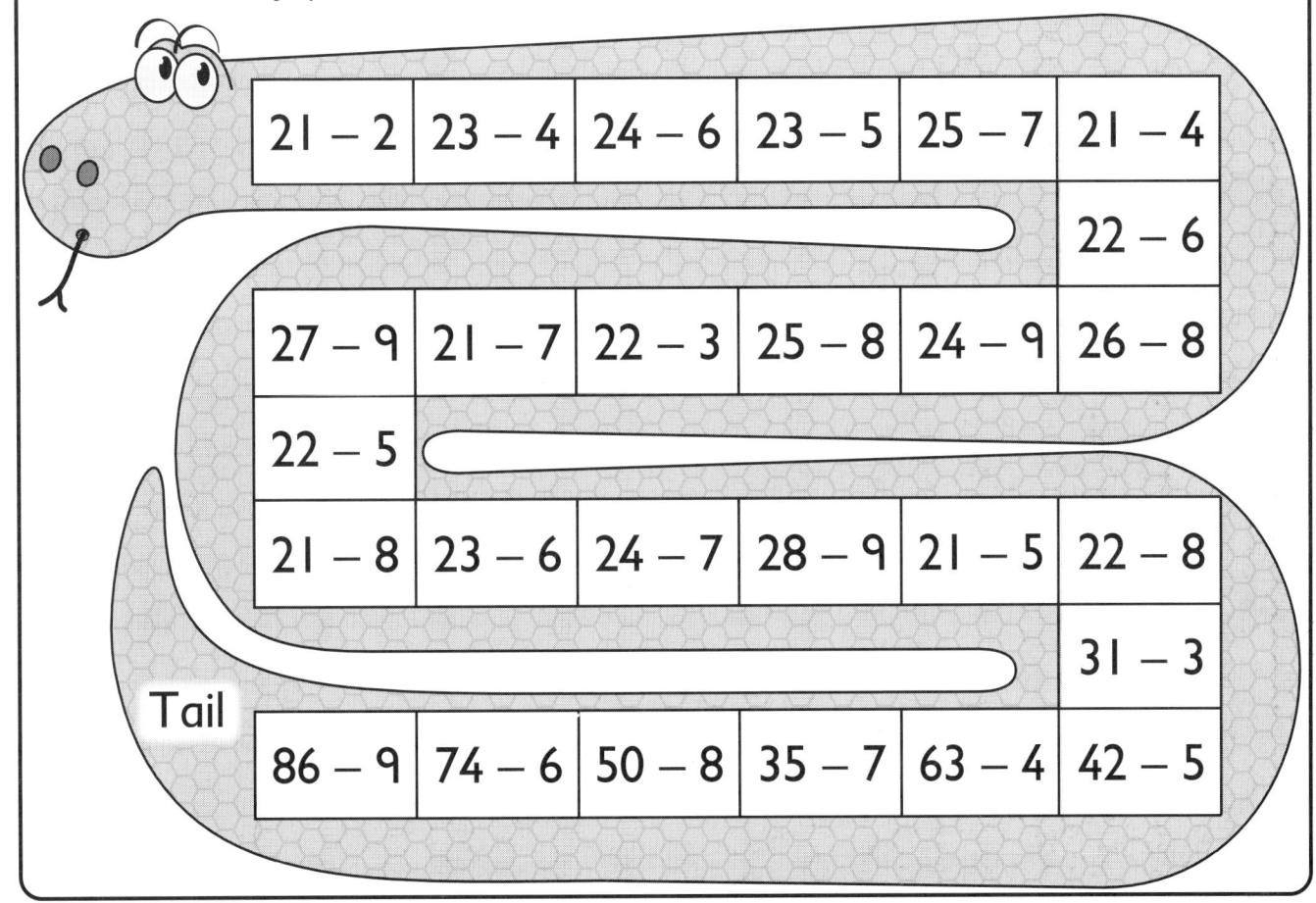

21 – 2	23 – 4	24 – 6	23 – 5	25 – 7	21 – 4
					22 – 6
27 – 9	21 – 7	22 – 3	25 – 8	24 – 9	26 – 8
22 – 5					
21 – 8	23 – 6	24 – 7	28 – 9	21 – 5	22 – 8
					31 – 3
86 – 9	74 – 6	50 – 8	35 – 7	63 – 4	42 – 5

Tail

---

21 – 3 = ☐   22 – 9 = ☐   23 – ☐ = 16

25 minus 6 = ☐   Subtract 8 from 24. ☐

91 – 2 = ☐   30 – 4 = ☐   53 – ☐ = 46

Subtraction to 100: Textbook page 28

Home Activity 11

Name:

Subtracting a teens number

- Ask the child to:
  - choose one car number from each row
  - write a subtraction story
  - cross out the numbers used
  - write the answer.
- Repeat for the remaining car numbers.

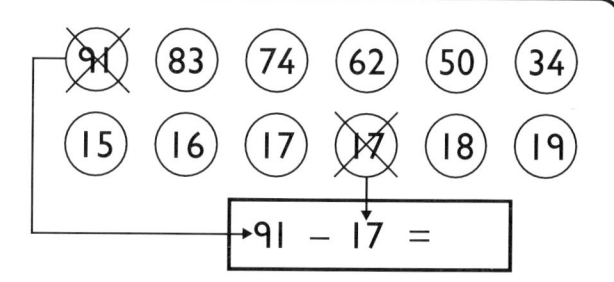

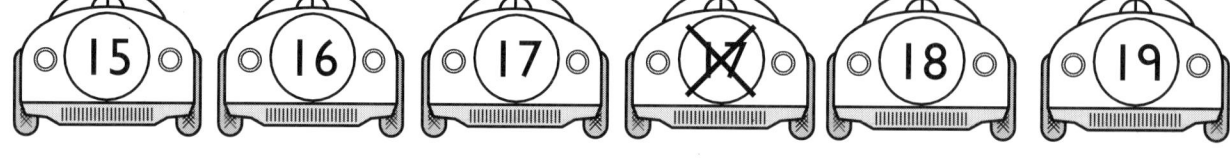

(⊗91) (83) (74) (62) (50) (34)
(15) (16) (17) (⊗17) (18) (19)

91 – 17 =

Cars row 1: ⊗91, 83, 74, 62, 50, 34
Cars row 2: 15, 16, 17, ⊗17, 18, 19

| 91 – 17 = | – = | – = |
| – = | – = | – = |

Help — Remind the child to subtract mentally in two steps. For 53 – 16:
step one 53 take away 10 ⟶ 43   *step two* 43 take away 6 ⟶ 37

42 – 13 = ☐   87 – 18 = ☐   95 – 17 = ☐

31 take away 15 = ☐   70 subtract 16 = ☐

63 – ☐ = 49   46 – ☐ = 28   ☐ – 16 = 39

Subtraction to 100: Textbook page 29

Two, five and ten times tables

Home Activity 12

- Ask the child to say the 'stations' of the two, five and ten times tables, forwards and backwards.

> Zero, five, ten, fifteen, twenty, ... fifty.

> One hundred, ninety, eighty, seventy, sixty, ... zero.

- Say part of a sequence from a table and have the child give the next number.

> Ten, twelve, fourteen, sixteen...

> Eighteen.

> Thirty-five, thirty, twenty-five, twenty...

> Fifteen.

- Now ask these questions:

> Two times nine?

> Five threes?

> Four multiplied by ten?

> Two sevens?

> Ten times eight?

> Five times six?

> Ten zeros?

> Nine multiplied by five?

> Six twos?

2 × 8 = ___

5 × 5 = ___

10 × 3 = ___

2 × 10 = ___

8 × 5 = ___

5 × 0 = ___

10 tens = ___

5 twos = ___

5 fours = ___

2 threes = ___

5 ones = ___

5 tens = ___

Multiplication: Textbook page 39

Name:

Home Activity 13

Three times table

- Ask the child to say the three times table.
- Now ask questions like these:

  Three times three? *Seven multiplied by three?*

  Eight threes? *Three nines?*

- Say part of a sequence from the three times table and ask the child to give the next number.

  Three, six, nine, twelve ... *Fifteen.*

  Twenty-seven, twenty-four, twenty-one... *Eighteen.*

Match.

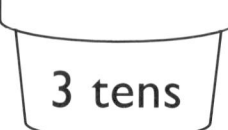

30   3 times 4   18   8 × 3

12   6 threes   3 tens   24

$3 \times \square = 15$ $\square \times 3 = 0$

$3 \times \square = 21$ $\square \times 3 = 6$

$3 \times \square = 3$ $\square \times 3 = 27$

Multiplication: Textbook page 40

Name:

Four times table

Home Activity 14

- Ask the child to say the four times table.
- Now ask questions like these:

  - Four times nine?
  - Four multiplied by four?
  - Four tens?
  - What are four sevens?
  - Four times what makes twenty?

- Point to each number on the grid below, in turn, and have the child say the appropriate fact from the four times table.

  For example:

0	4	12
6	32	28

  Four times three makes twelve.

8	24	40	4	12
20	16	36	32	28

Mark each example ✓ or ✗.

4 X 9 = 36   4 X 8 = 32   4 X 1 = 1

4 X 6 = 24   4 X 0 = 4   4 X 7 = 28

4 X 10 = 30   4 X 2 = 8   4 X 3 = 12

Multiplication: Textbook page 41

Multiplying by 2, 3, 4, 5 and 10

Find the answers. Colour them on the grid below to read the alien's message.

$3 \times 6 =$ ☐　　$5 \times 5 =$ ☐

$10 \times 2 =$ ☐　　$4 \times 8 =$ ☐

$9 \times 3 =$ ☐　　$10 \times 4 =$ ☐　　$4 \times 4 =$ ☐

$3 \times 7 =$ ☐　　$4 \times 9 =$ ☐　　$4 \times 3 =$ ☐

$5 \times 3 =$ ☐　　$4 \times 7 =$ ☐　　$3 \times 10 =$ ☐

32	45	24	22	28	3
15	4	10	35	33	70
21	18	30	5	40	6
20	9	16	14	36	0
12	2	27	50	25	8

Home Activity 15

Multiplication: Textbook page 43

Multiplying two-digit numbers

Home Activity 16

What is the cost of three footballers?

One hundred and twenty pence.

Ask for the cost of:
- 2 footballers
- 3 teachers
- 2 teachers
- 10 doctors
- 4 doctors
- 5 footballers.

TROLLS

- Teacher 30p
- Footballer 40p
- Doctor 20p
- Cook 35p
- Police Officer 12p
- Artist 23p

Find the cost of

2 (Police Officer) ☐ 2 (Cook) ☐ 2 (Artist) ☐

3 (Artist) ☐ 3 (Footballer) ☐ 4 (Police Officer) ☐

Multiplication: Textbook page 45

Home Activity 17

Dividing by 2 and 10

Twenty divided by two?

Ten.

Correct. Colour a kite with ten on it.

Repeat for these examples:
- 16 divided by 2
- 40 divided by 10
- 80 divided by 10
- 10 divided by 2
- 8 divided by 2
- 100 divided by 10

14 ÷ 2 = ____   6 ÷ 2 = ____   18 ÷ 2 = ____

90 ÷ 10 = ____   50 ÷ 10 = ____   10 ÷ 10 = ____

half of 4 = ____   half of 60 = ____

How many twos make 12? ____

How many tens make 70? ____

Division: Textbook page 48

Dividing by 5

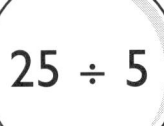

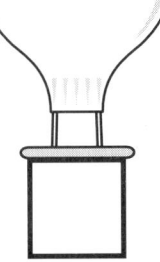

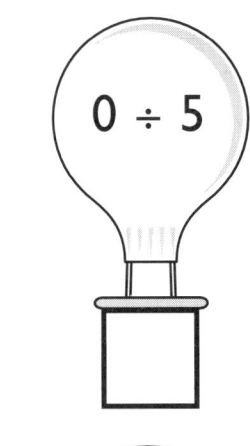

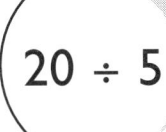

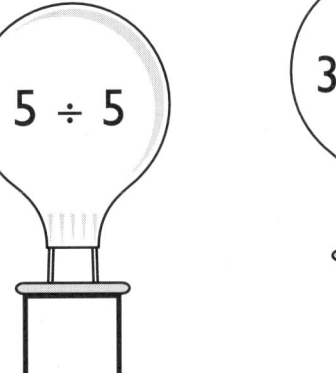

- 25 ÷ 5
- 50 ÷ 5
- 10 ÷ 5
- 0 ÷ 5
- 20 ÷ 5
- 40 ÷ 5
- 5 ÷ 5
- 35 ÷ 5

Help Allow the child to jump back in fives on the number line.

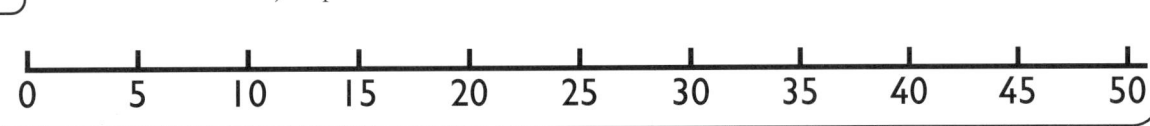

0 5 10 15 20 25 30 35 40 45 50

Divide 45 by 5. ☐   15 divided by 5 = ☐

How many fives make 30? ☐

Share 50 equally among 5. ☐

How many groups of five make 25? ☐

Home Activity 19

Dividing by 3

Use a  🎲 and place a counter for each player at the Start.
In turn, each player:
- rolls the die to find how many boxes along the track to move their finger
- divides the number in that box by 3
- if correct, moves their counter to the box
- if wrong, does not move.

The first player to reach the Finish wins.

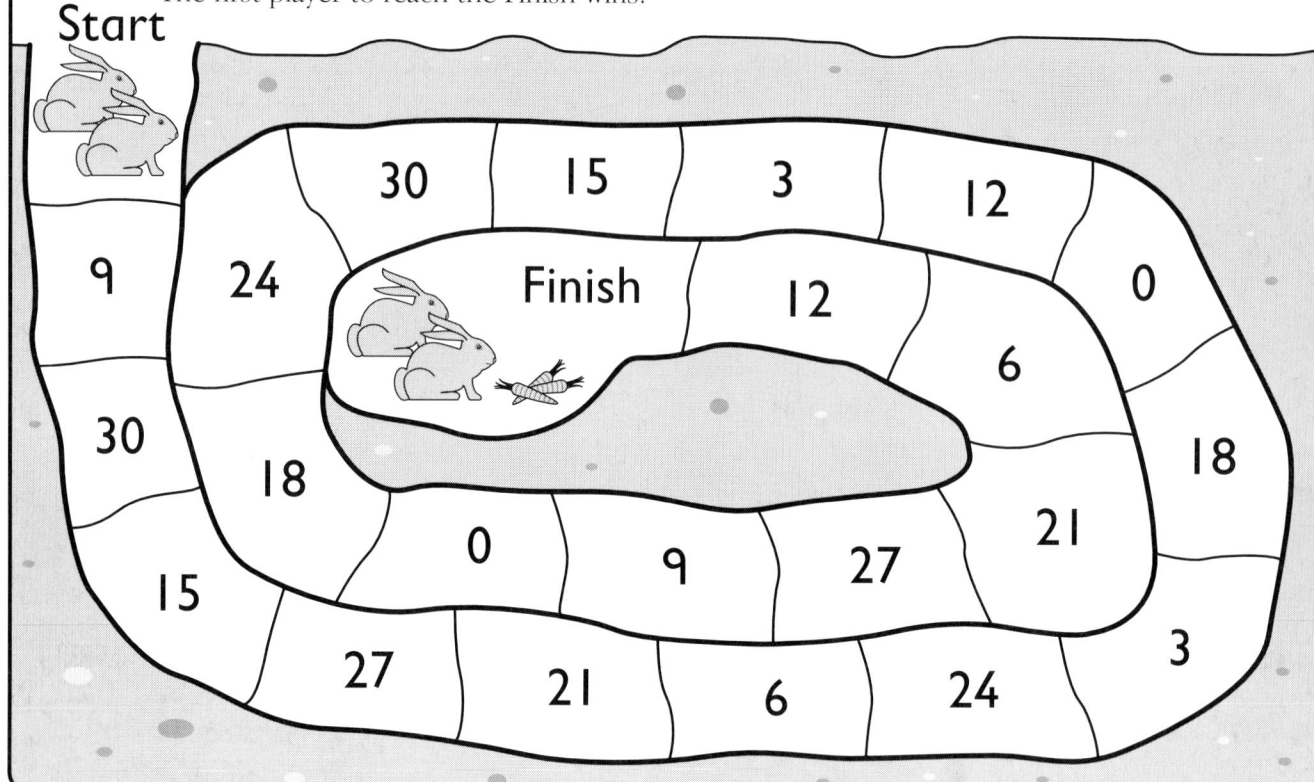

Start

Board track numbers: 9, 30, 15, 27, 21, 6, 24, 3, 18, 27, 9, 0, 18, 24, 30, 15, 3, 12, Finish, 12, 6, 21, 0

Divide 30 equally among 3. ☐   15 divided by 3 ☐

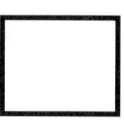

How many threes make 24? ☐   Divide 3 by 3. ☐

12 ÷ 3 = ☐   21 ÷ 3 = ☐   9 ÷ 3 = ☐

☐ ÷ 3 = 2   ☐ ÷ 3 = 9   ☐ ÷ 3 = 6

Division: Textbook page 52

Name:

Dividing by 4

Home Activity 20

- Ask the child to:
  - choose a number on the track
  - divide the number by 4
  - colour the box the chosen number is in.
- Repeat until all the boxes have been coloured.

| 8 | 24 | 36 | 4 | 32 | 12 | 20 | 28 | 0 | 40 | 16 |

Divide 20 by 4. _____

32 divided by 4. _____

One quarter of 16 _____

36 ÷ 4 = _____

4 ÷ 4 = _____

40 ÷ 4 = _____

24 ÷ 4 = _____

Colour answers red.

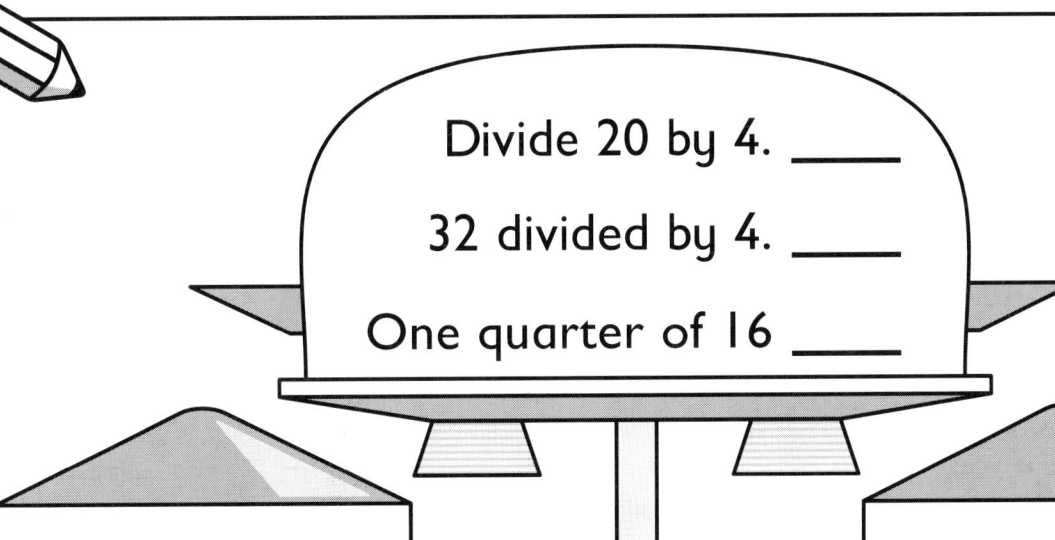

Colour answers blue.

28 ÷ 4 = _____

0 ÷ 4 = _____

12 ÷ 4 = _____

8 ÷ 4 = _____

Colour answers green.

8   0   2   10   5   9

7   1   4   6   3

Division: Textbook page 54

Dividing by 2, 3, 4, 5 and 10

Scatter the division cards face down on the table.
Each player, in turn:
– chooses two cards to turn face up
– answers the division on each card
– if the answers are the same, keeps the pair of cards;
  if the answers are different, replaces the cards face down.
The player with most pairs at the end wins.

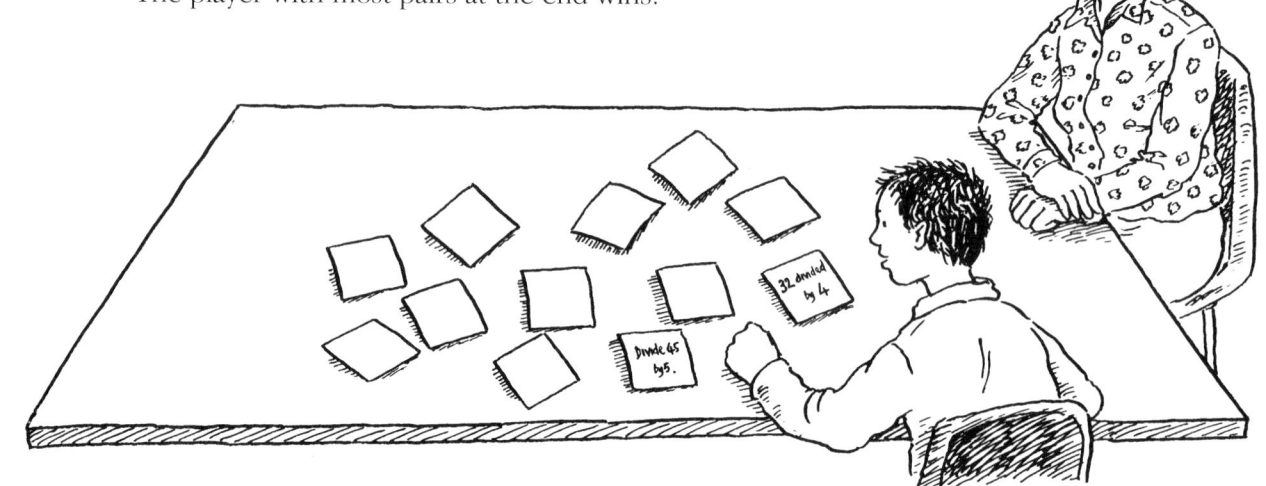

Share 35 equally among 5. ☐   16 divided by 2 ☐

How many tens make 60? ☐   Divide 24 by 4. ☐

One quarter of 36 ☐   Half of 14 ☐

20 ÷ 5 = ☐   18 ÷ 3 = ☐   100 ÷ 10 = ☐

0 ÷ 4 = ☐   30 ÷ 5 = ☐   15 ÷ 3 = ☐

☐ ÷ 10 = 4   ☐ ÷ 3 = 10   ☐ ÷ 2 = 9

Division: Textbook page 56

Home Activity 22

Name:

Money

- Use 1p, 2p, 5p, 10p, 20p, 50p, £1 and £2 coins.
- Show the child collections of coins and ask for the total amount each time.

"Three pounds and fifty-four pence."

- Ask the child to lay out coins to show specific amounts.

"Show me two pounds and eighty-five pence."

Help Encourage the child to lay out as few coins as possible, starting with the largest value.

Write each amount in **pounds** and **pence**.

257p   303p   85p

_____   _____   _____

Write each amount in **pence**.

£3·20   £4·06   £0·68

_____   _____   _____

How much does Tim have left?

"I had £2. I spent 90p."

_____

Money: Textbook page 64

£5, £10 and £20 notes

Home Activity 23

How much?

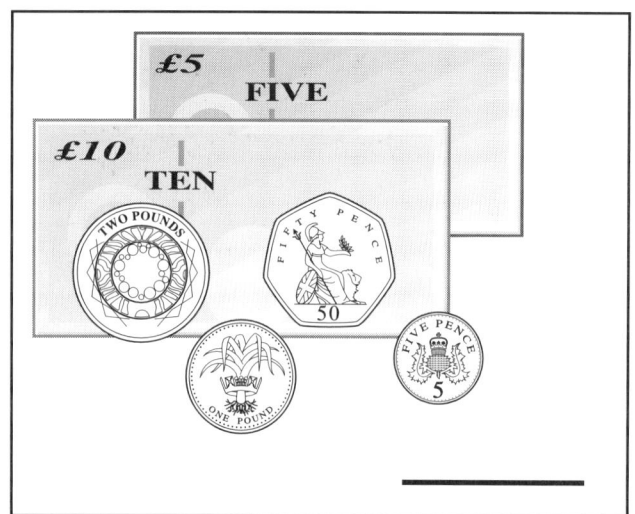

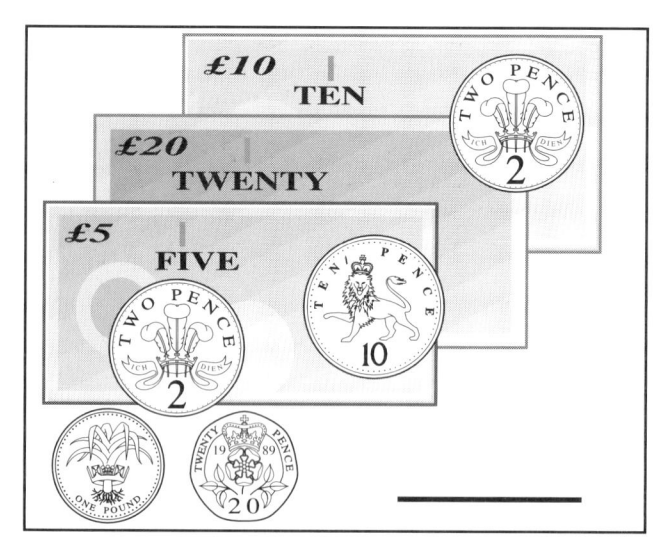

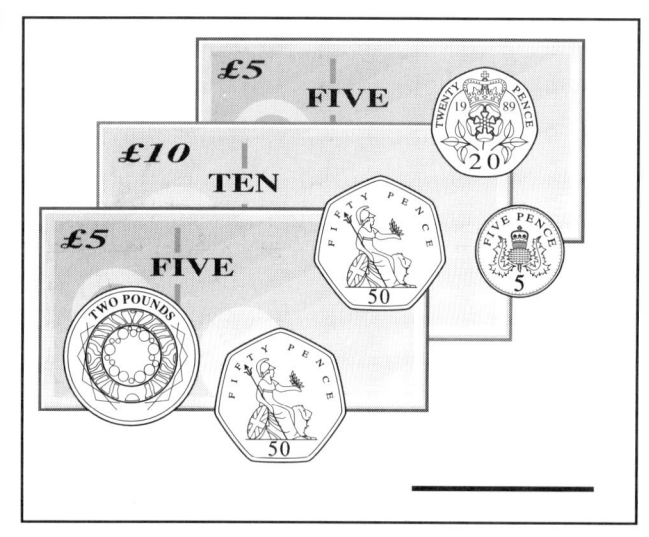

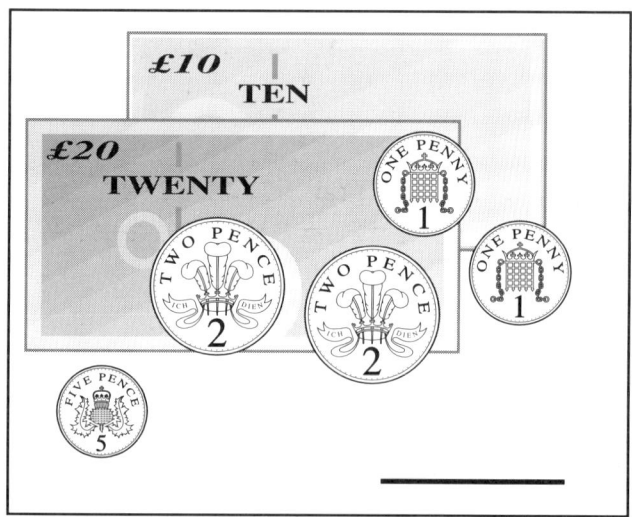

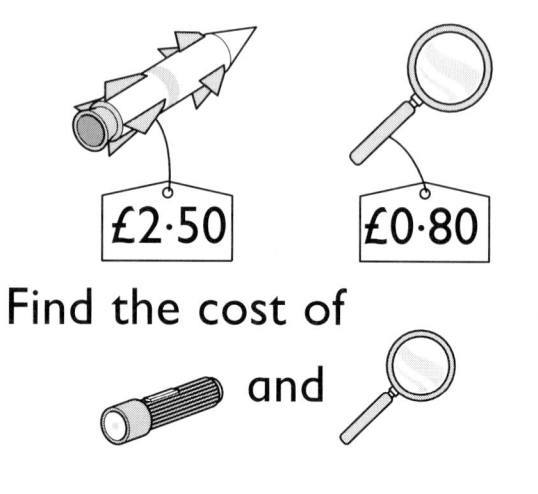

1. £10 + £5 + £2 + 50p + £1 + 5p = £18.55

2. £20 + £10 + £5 + £1 + 10p + 2p + 20p + 2p = £36.34

3. £10 + £5 + £5 + 50p + 50p + 20p + 5p = £21.25

4. £20 + £10 + £2 + £2 + 5p + 1p + 1p = £34.07

Prices:
- rocket £2·50
- magnifying glass £0·80
- torch £1·20
- radio £3·30

Find the cost of torch and magnifying glass _____

Find the cost of radio and rocket _____

Money: Textbook page 67

Tenths, '$\frac{1}{10}$ of'

Home Activity 24

- Ask the child to: *Colour four tenths of the square.*
- Repeat for: six tenths of the circle
  nine tenths of the rectangle
  two tenths of the pentagon.

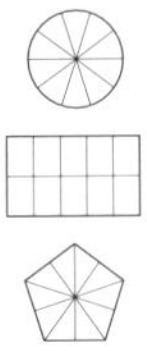

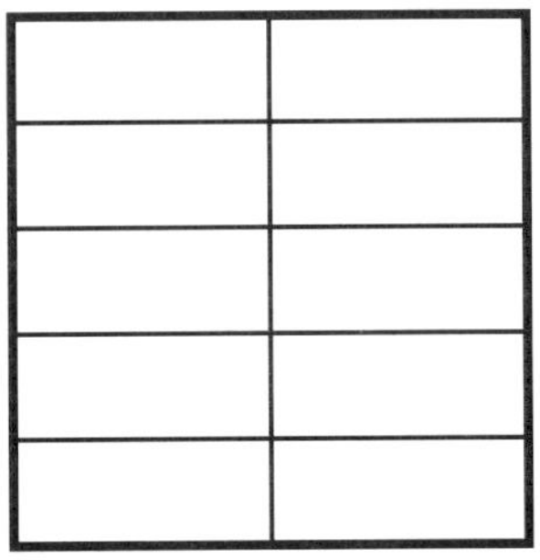

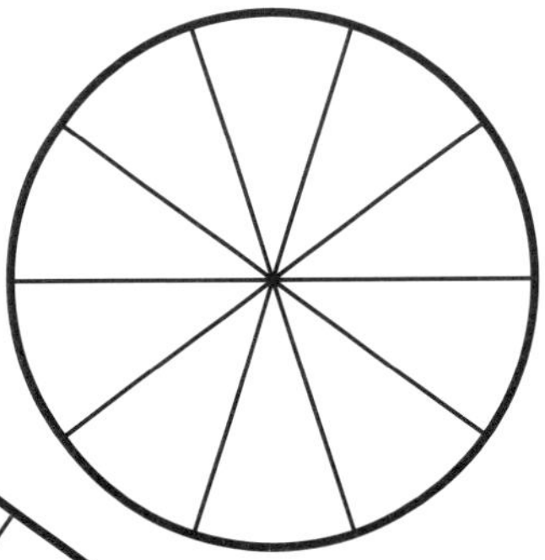

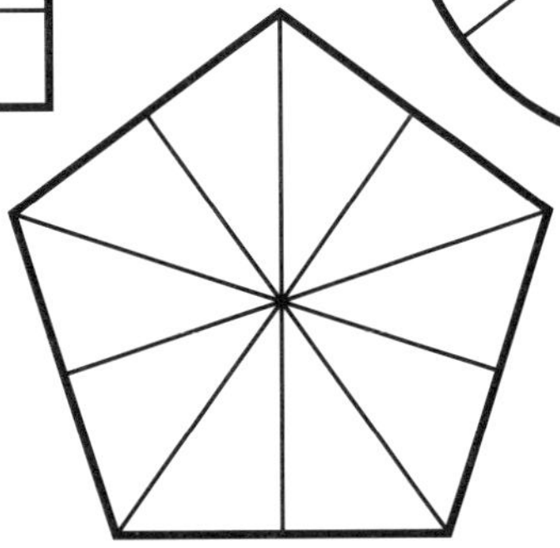

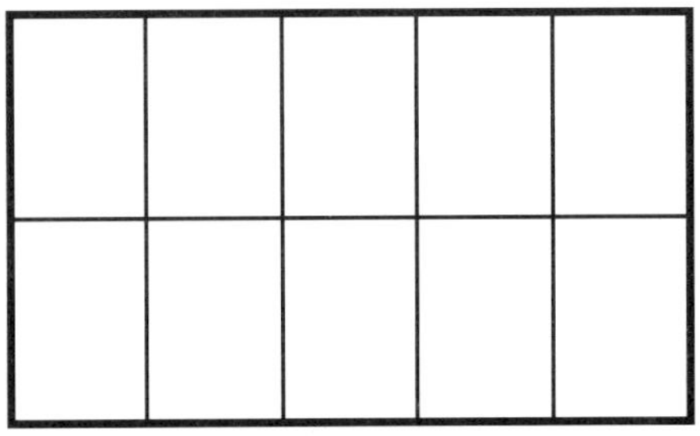

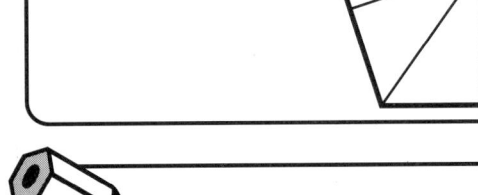

one tenth of 50 = ☐ $\frac{1}{10}$ of 20 = ☐

one tenth of 80 = ☐ $\frac{1}{10}$ of 100 = ☐

one tenth of 10 = ☐ $\frac{1}{10}$ of 70 = ☐

There are 90 counters in the bag.
One tenth of them are red. ☐
How many are **not** red?

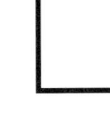

Help: To find one tenth of, for example 50, the child could:
– divide 50 by 10 or
– think 'Ten times what make fifty?'

Fractions: Textbook page 72

Home Activity 25

Name: _____

Thirds, fifths

Colour

one third

two thirds

three thirds

one third red
two thirds blue.

What fraction is **not** shaded? _____

Tick (✓) the shapes which show fifths.

Colour

three fifths red
two fifths yellow

one fifth red
four fifths blue.

What fraction is

• shaded _____

• not shaded? _____

Fractions: Activity Book page 32

Home Activity 26

Writing times

Write each time using **minutes past** or **minutes to**.

_____   _____

The time on this clock is **4.35**

Write each of these times.

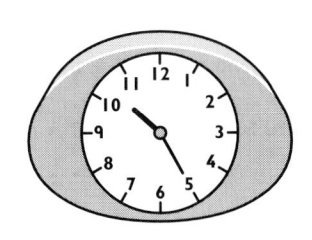

_____   _____   _____

Quarter past 8.   10 minutes to 5.

_____   _____

20 minutes past 7.   Quarter to 11.

_____   _____

Time: Textbook page 102

Home Activity 27

Time intervals

How long is it from five minutes past two until twenty minutes past two?

Five, ten, fifteen minutes.

- Repeat for these time intervals:

  15 minutes past 7 until 25 minutes to 8   25 minutes to 3 until 3 o'clock
  20 minutes past 5 until 10 minutes to 6   10 minutes to 12 until 10 minutes past 12.

- Now ask for these time intervals:

  6.30 until 9.30   2.05 until 4.05   11.15 until 3.15

Help Use the clock face. The child should count on in five minute intervals or one hour intervals from the starting time.

How long?

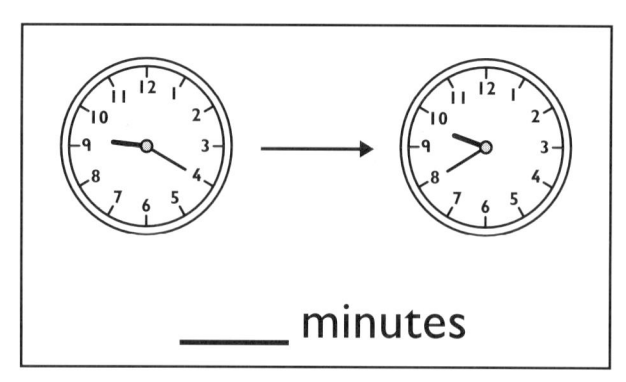

____ minutes

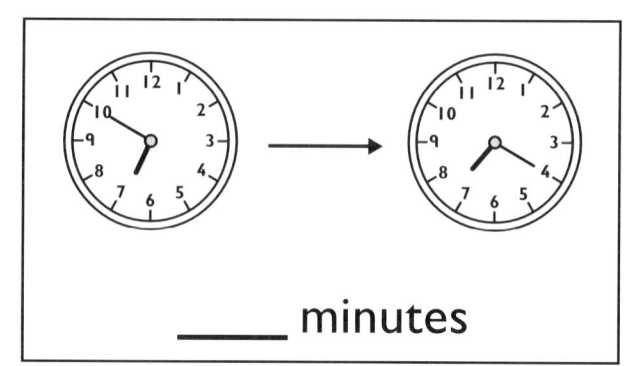

____ minutes

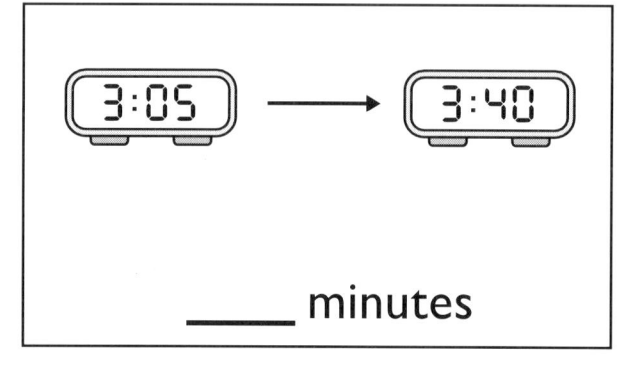

3:05 → 3:40

____ minutes

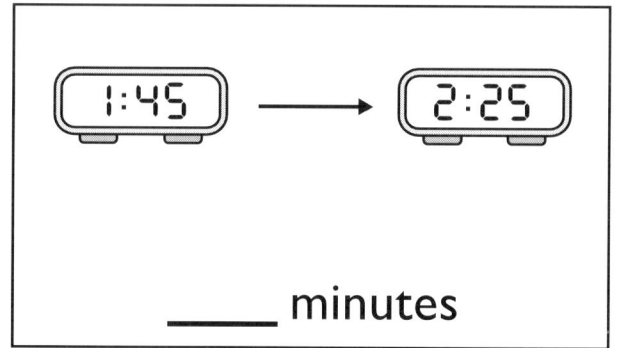

1:45 → 2:25

____ minutes

Time: Activity Book page 36

Home Sheet 28

Name:

Domino cards for Home Activity 2

805	10 more than 250	863	480 add 20
706	100 more than 900	676	855 minus 50
500	349 add 40	260	100 less than 806
1000	10 less than 873	389	696 minus 20

✂ Cut out these domino cards.

Name:

Home Sheet 29

Number cards for Home Activities 3 and 8

1	2	3
4	5	6
7	8	9

✂ Cut out these number cards.

Name:

Home Sheet 30

Number cards for Home Activity 8

10	11	12
13	14	15
16	17	18
19	20	0

✂ Cut out these number cards.

Name:

Home Sheet 31

Division cards for Home Activity 21

How many twos make 8?	One quarter of 16	15 shared equally among 3
Divide 50 by 10.	28 divided by 4	Divide 21 equally among 3.
32 divided by 4	How many groups of 10 make 80?	Divide 45 by 5.
How many threes make twenty-seven?	Share 50 equally among 5.	Share 20 equally between 2.

✂ Cut out these division cards.

Resource Sheets

Resource Sheet 1	Number grid template
Resource Sheet 2	Number grid overlay
Resource Sheet 3	Number machine template
Resource Sheet 4	Connection cards
Resource Sheet 5	Connection cards
Resource Sheet 6	Connection cards
Resource Sheet 7	Connection cards
Resource Sheet 8	Flashcards (number after/before/between)
Resource Sheet 9	Flashcards (one more than/one less than/1 more than)
Resource Sheet 10	Flashcards (1 less than/two more than/two less than)
Resource Sheet 11	Flashcards (2 more than/2 less than/ten more than)
Resource Sheet 12	Flashcards (ten less than/10 more than/10 less than)
Resource Sheet 13	Flashcards (100 more than/100 less than/… less than)
Resource Sheet 14	Flashcards (take … from/subtraction/minus)
Resource Sheet 15	Flashcards (take away/subtract … from/difference between … and)
Resource Sheet 16	Flashcards (more than/add/plus)
Resource Sheet 17	Flashcards (*What* add … makes/… add … and/add *what* makes)
Resource Sheet 18	Flashcards (times/multiply … by/… multiplied by)
Resource Sheet 19	Flashcards (double/twice/find the product of … and)
Resource Sheet 20	Flashcards (divide … by/ … divided by/share … equally among)
Resource Sheet 21	Flashcards (divide equally … between/How many … in/group … in)
Resource Sheet 22	Flashcards (one half of/one quarter of/one tenth of)
Resource Sheet 23	Flashcards (o'clock/half past/quarter past/quarter to
Resource Sheet 24	Months (January/February/March/April)
Resource Sheet 25	Months (May/June/July/August)
Resource Sheet 26	Months (September/October/November/December)
Resource Sheet 27	Shape (North, South, East, West)
Resource Sheet 28	Shape (Forward/Back/Right/Left)
Resource Sheet 29	Number cards 0–14
Resource Sheet 30	Number cards 15–29
Resource Sheet 31	Number cards 30–44
Resource Sheet 32	Number cards 45–59
Resource Sheet 33	Number cards 60–74
Resource Sheet 34	Number cards 75–89
Resource Sheet 35	Number cards 90–100
Resource Sheet 36	Place value cards
Resource Sheet 37	Place value cards
Resource Sheet 38	Place value cards
Resource Sheet 39	Number square 1–100
Resource Sheet 40	Number square 1–100

Resource Sheet 41 Ordinal number cards (1st – 8th)
Resource Sheet 42 Ordinal number cards (9th – 16th)
Resource Sheet 43 Ordinal number cards (17th – 20th) number names (first – second)
Resource Sheet 44 Ordinal number names (third – sixth)
Resource Sheet 45 Ordinal number names (seventh – tenth)
Resource Sheet 46 Ordinal number names (eleventh – fourteenth)
Resource Sheet 47 Ordinal number names (fifteenth – eighteenth)
Resource Sheet 48 Ordinal number names (nineteenth/twentieth/fiftieth/hundredth)
Resource Sheet 49 Ten frames
Resource Sheet 50 Time cards (am ⟶ pm times)
Resource Sheet 51 Time cards (am ⟶ pm times)
Resource Sheet 52 Time cards (am ⟶ pm times)
Resource Sheet 53 Isometric dot grid

Number grid template

Number grid overlay

NHM 3 New Heinemann Maths 2

Photocopy onto card then cut out the 4 shaded squares to create a card overlay for the number grid on Resource Sheet 1.

New Heinemann Maths 3 • © SPMG 2000 • Copying permitted for purchasing school only. This material is not copyright free.

Calculation grid template

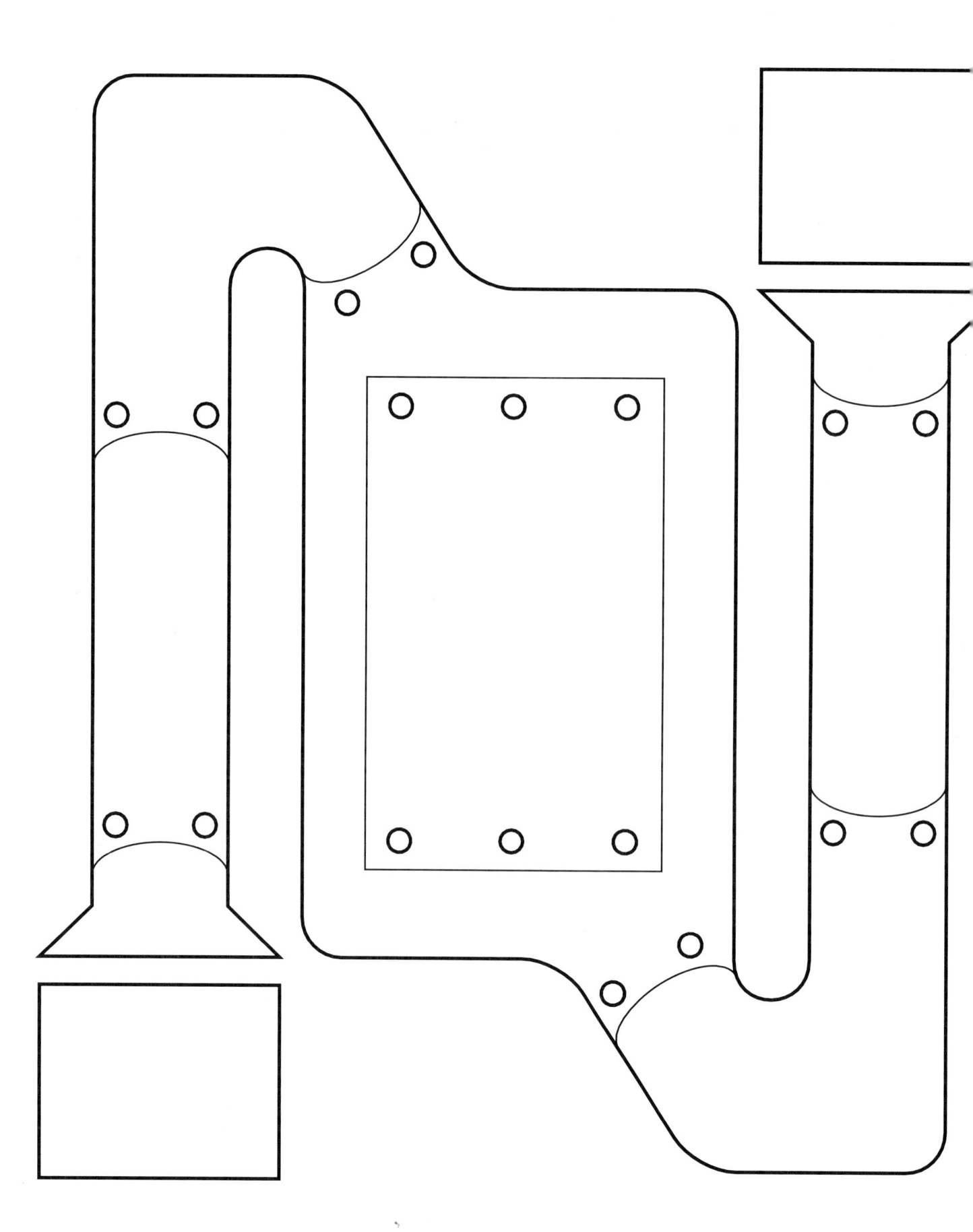

Connection cards

NHM 3 — 4

25	10 × 10	100	5 times 9
45	Multiply 7 by 3.	21	10 × 8
80	3 times 4	12	3 × 6
18	Multiply 9 by 3.	27	10 times 5
50	5 times 1 add 2	7	2 × 7

Connection cards

14	Double 5	10	3 times 3
9	4 × 5	20	2 fours
8	10 times 6	60	Multiply 1 by 3.
3	5 times 3	15	Double 10 add 2
22	10 × 9	90	Double 2

Connection cards

4	Double 9, subtract 1	17	2 sets of 3
6	Multiply 6 by 4.	24	4 × 7
28	10 times 0	0	4 × 8
32	Multiply 3 by 10.	30	4 nines
36	10 × 7	70	4 multiplied by 4

New Heinemann Maths 3 • © SPMG 2000 • Copying permitted for purchasing school only. This material is not copyright free.

Connection cards

16	4 threes, plus 1	13	5 × 7
35	5 times 2, add 1	11	5 times 8, subtract 1
39	Multiply 4 by 10.	40	5 sixes, plus 1
31	3 times 8, subtract 1	23	10 twos, minus 1
19	3 tens, plus 3	33	5 × 5

Flashcards (number after/before/between)

the number after

the number before

and

between

Flashcards (one more than/one less than/1 more than)

one more than

one less than

1 more than

Flashcards (1 less than/two more than/two less than)

1 less than

two more than

two less than

Flashcards (2 more than/2 less than/ten more than)

2 more than

2 less than

ten more than

Flashcards (ten less than/10 more than/10 less than)

NHM 3 | 12

ten less than

10 more than

10 less than

Flashcards (100 more than/100 less than/...less than)

100 more than

100 less than

less than

Flashcards (take... from/subtract/minus)

NHM 3 14

take from

subtract

minus

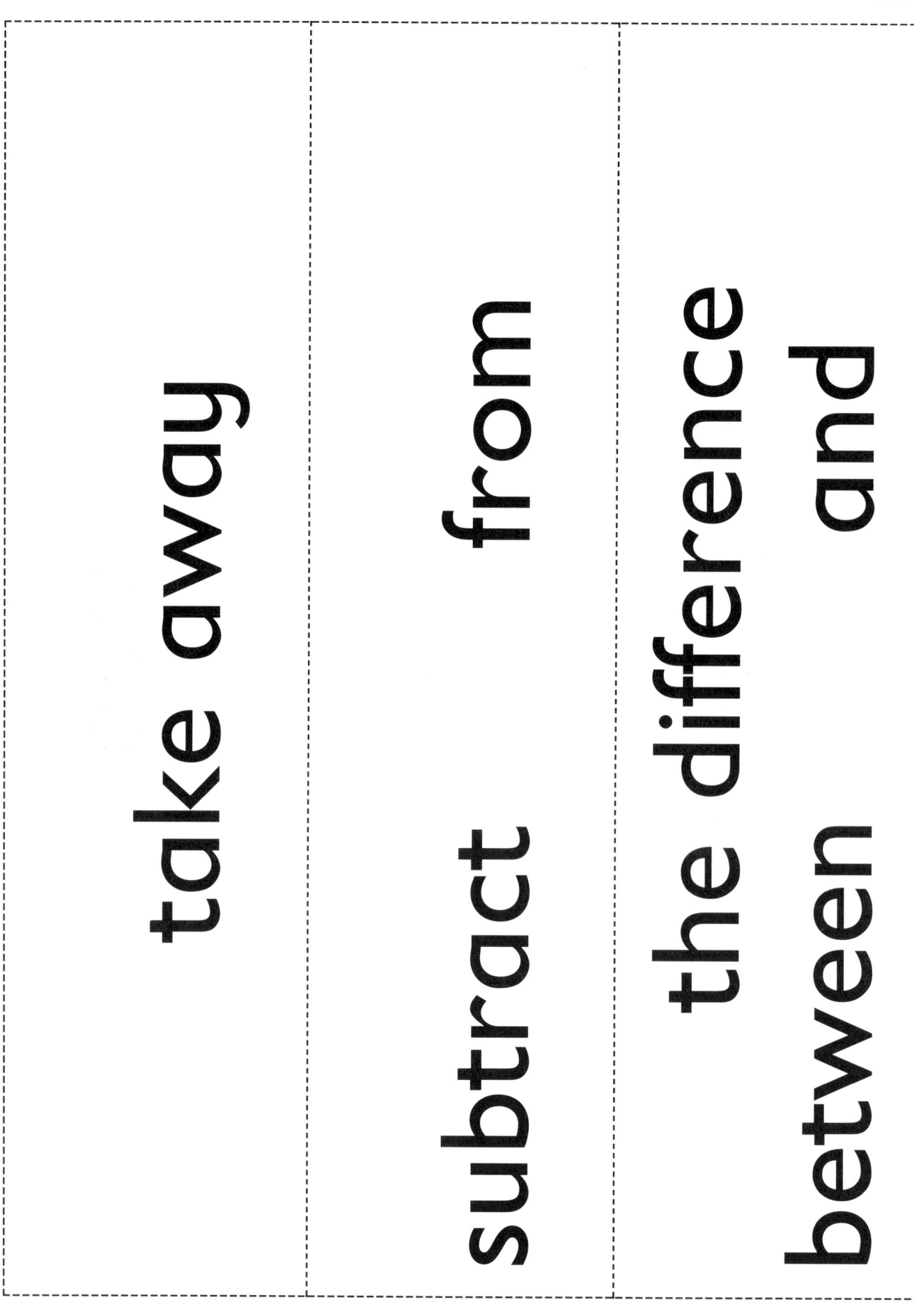

Flashcards (take away/subtract... from/ the difference between... and)

NHM 3 — 15

take away

subtract

from

the difference between

and

New Heinemann Maths 3 • © *SPMG 2000* • Copying permitted for purchasing school only. This material is not copyright free.

Flashcards (more than/add/plus)

more than

add

plus

Flashcards *(What add... makes/ ... add... and/add what makes)*

What add   makes

add   and

add what makes

Flashcards (times/multiply... by/... multiplied by)

NHM 3 — 18

times

multiply by

multiplied by

Flashcards (double/twice/find the product of... and)

NHM 3 — 19

double

twice

find the product of ... and

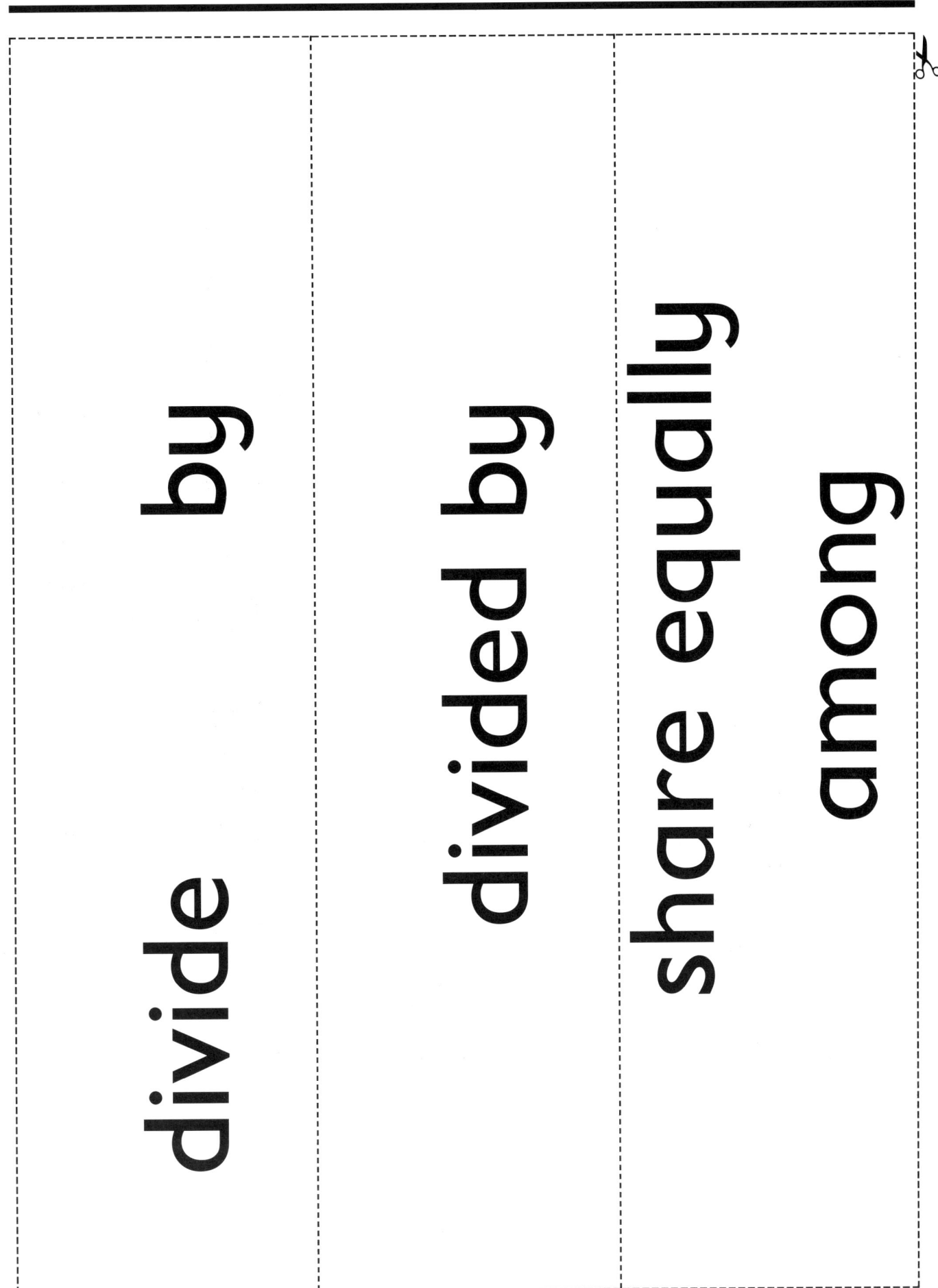

Flashcards (divide... by/... divided by/ share equally... among)

NHM 3 — 20

divide by

divided by

share equally among

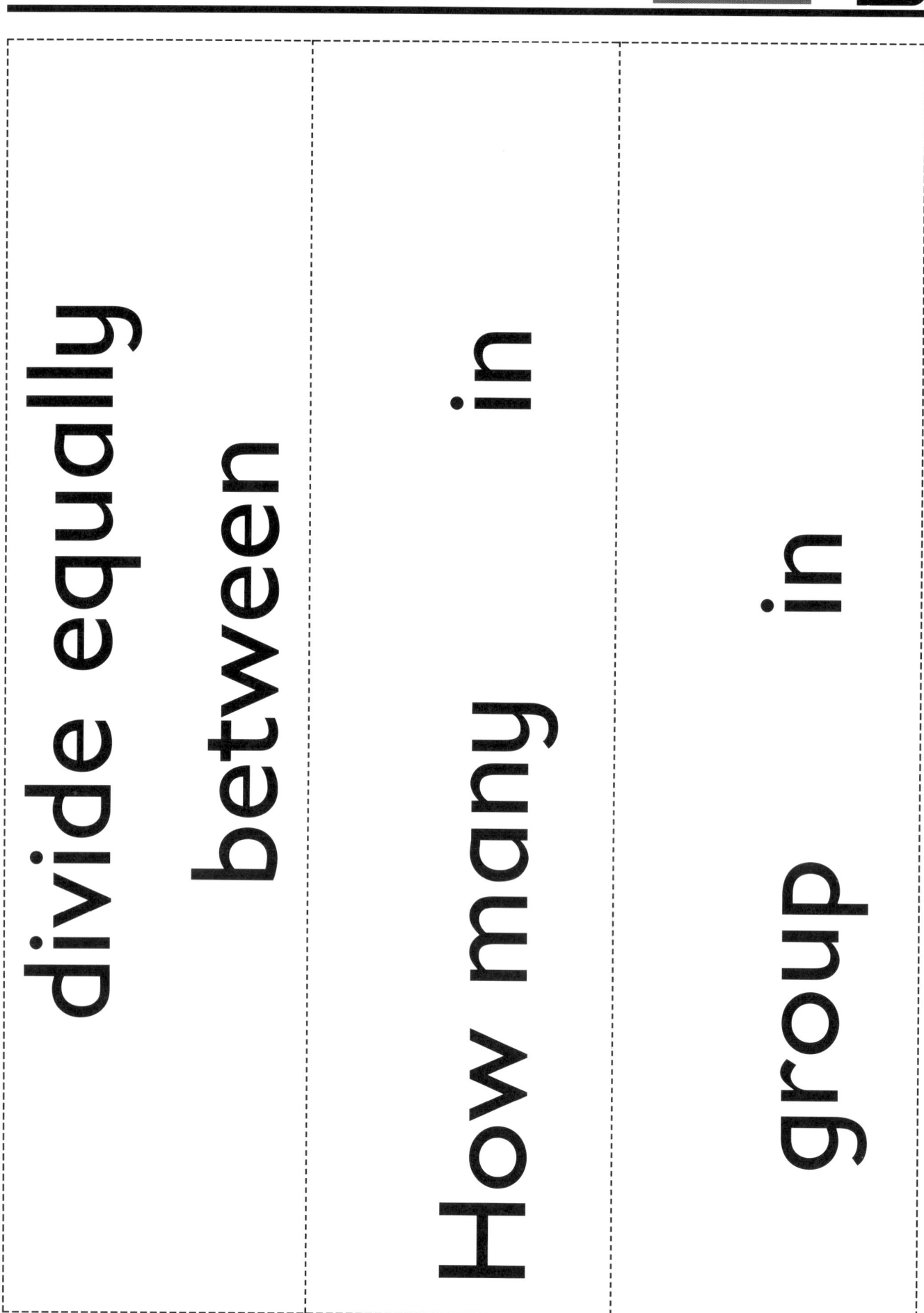

Flashcards

(divide equally... between/ How many... in/group... in)

NHM 3 — 21

divide equally... between

How many... in

group... in

Flashcards (one half of/one quarter of/one tenth of)

one half of

one quarter of

one tenth of

Flashcards (o'clock/half past/quarter past/quarter to)

o'clock

half past

quarter past

quarter to

Months (January/February/March/April)

NHM 3 — 24

January

February

March

April

Months (May/June/July/August)

May

June

July

August

Months (September/October/November/December)

September

October

November

December

Shape (North, South, East, West)

North

South

East

West

Shape (Forward/Back/Right/Left)

NHM 3 28

Forward

Back

Right

Left

Number cards 0–14

4	9	14
3	8	13
2	7	12
1	6	11
0	5	10

New Heinemann Maths 3 • © SPMG 2000 • Copying permitted for purchasing school only. This material is not copyright free.

Number cards 15–29

19	24	29
18	23	28
17	22	27
16	21	26
15	20	25

Number cards 30–44

30	31	32	33	34
35	36	37	38	39
40	41	42	43	44

Number cards 45–59

45	46	47	48	49
50	51	52	53	54
55	56	57	58	59

Number cards 60–74

60	61	62	63	64
65	66	67	68	69
70	71	72	73	74

Number cards 75–89

75	80	85
76	81	86
77	82	87
78	83	88
79	84	89

Number cards 90–100

94	99
93	98
92	97
91	96
90	95	100

Place value cards

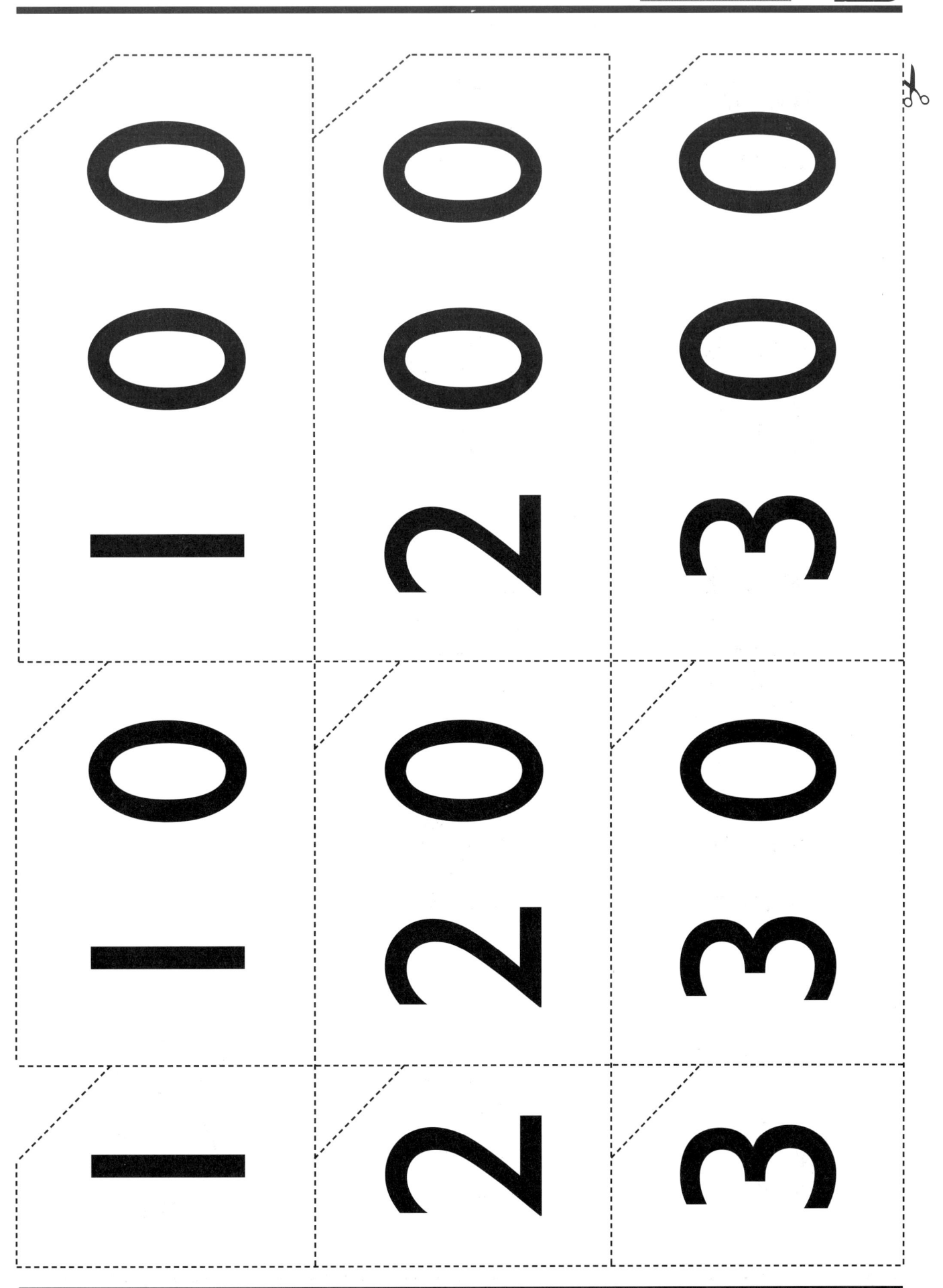

36

100	200	300
10	20	30
1	2	3

New Heinemann Maths 3 • © SPMG 2000 • Copying permitted for purchasing school only. This material is not copyright free.

Place value cards

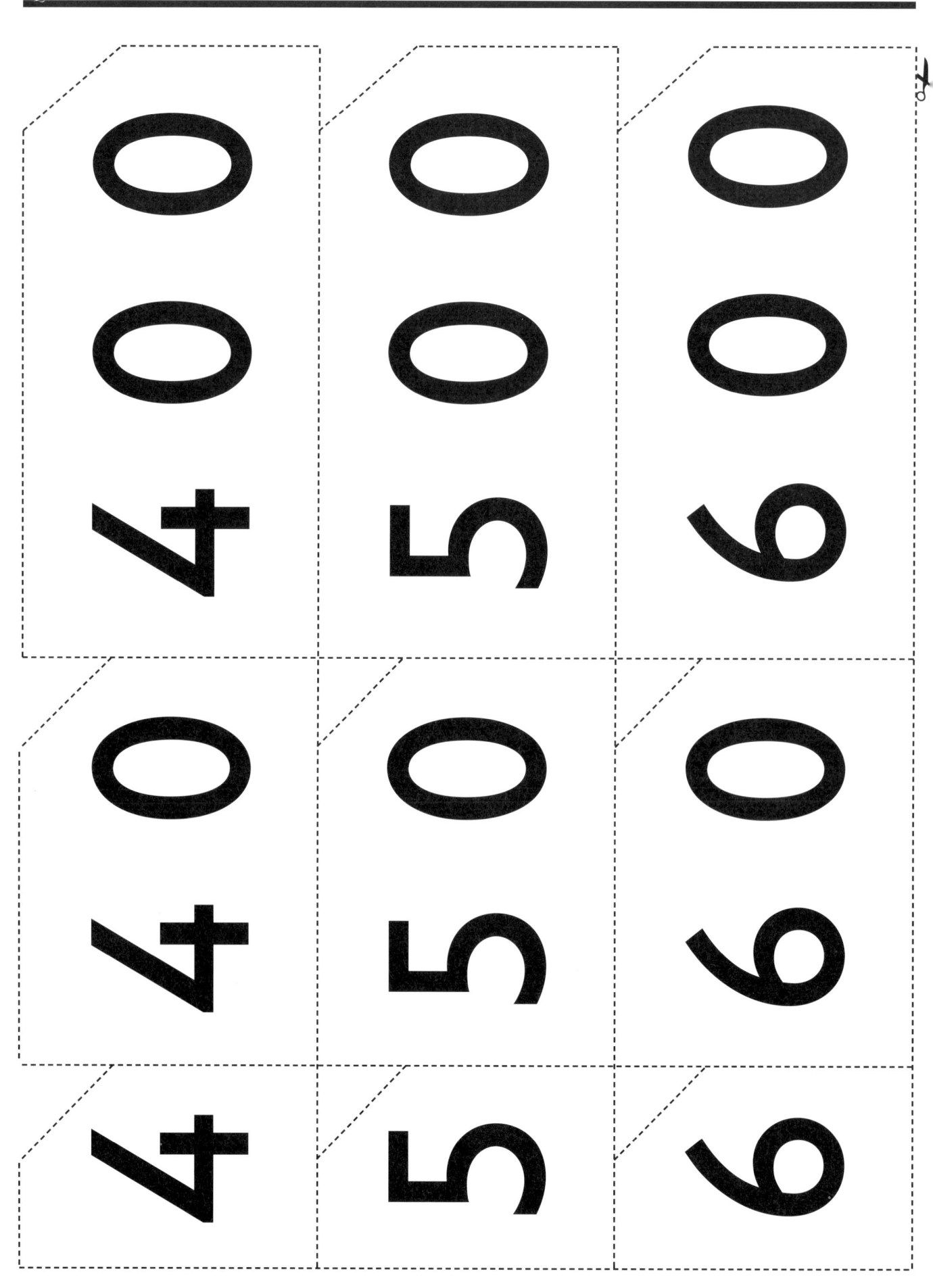

NHM 3 37

4000	5000	6000
400	500	600
4	5	6

Place value cards

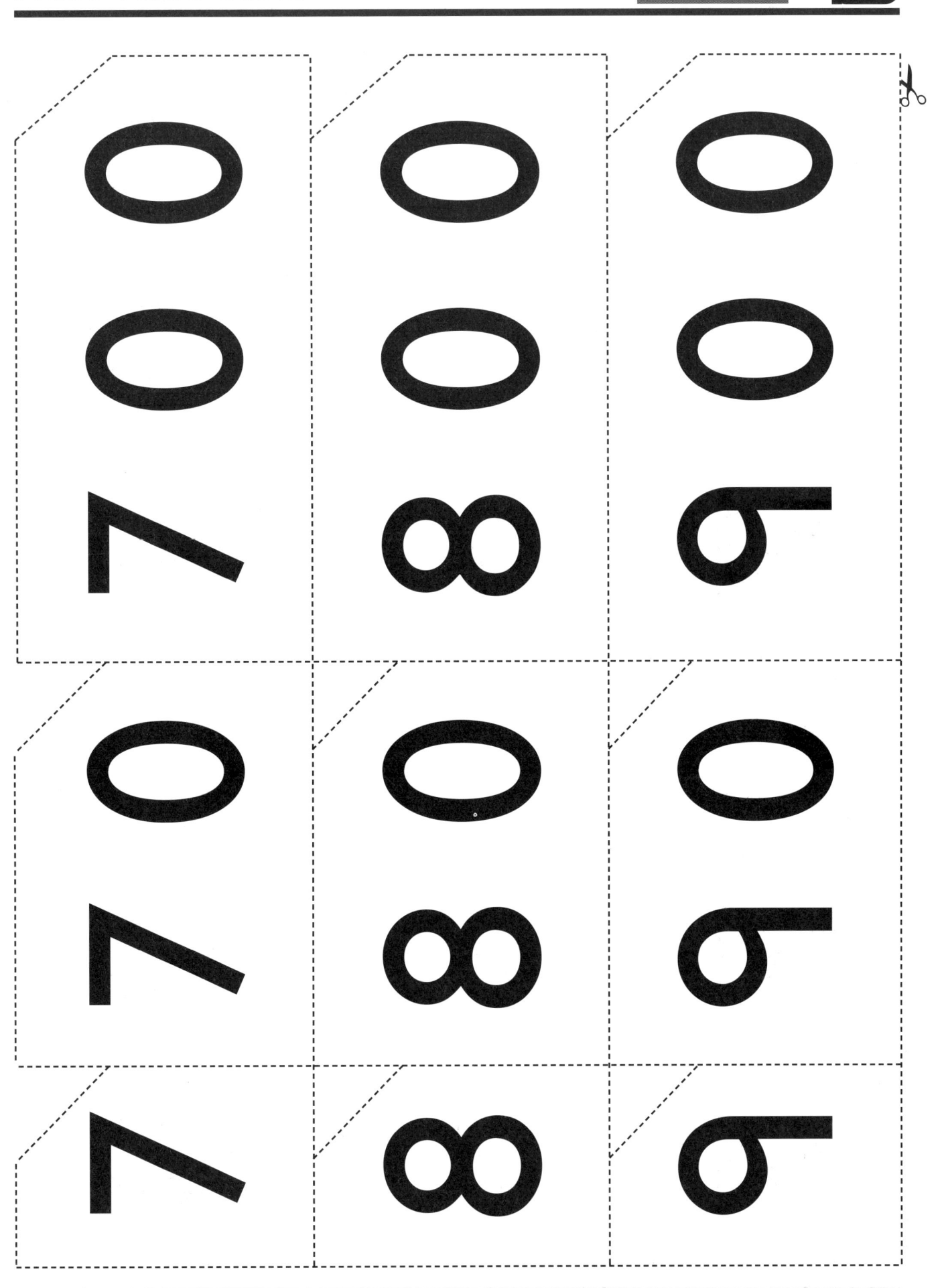

NHM 3 — 38

Number square 1–100

1	2	3	4	5	6	7	8	9	10
11	12	13	14	15	16	17	18	19	20
21	22	23	24	25	26	27	28	29	30
31	32	33	34	35	36	37	38	39	40
41	42	43	44	45	46	47	48	49	50
51	52	53	54	55	56	57	58	59	60
61	62	63	64	65	66	67	68	69	70
71	72	73	74	75	76	77	78	79	80
81	82	83	84	85	86	87	88	89	90
91	92	93	94	95	96	97	98	99	100

Number square 1–100

91	92	93	94	95	96	97	98	99	100
81	82	83	84	85	86	87	88	89	90
71	72	73	74	75	76	77	78	79	80
61	62	63	64	65	66	67	68	69	70
51	52	53	54	55	56	57	58	59	60
41	42	43	44	45	46	47	48	49	50
31	32	33	34	35	36	37	38	39	40
21	22	23	24	25	26	27	28	29	30
11	12	13	14	15	16	17	18	19	20
1	2	3	4	5	6	7	8	9	10

Ordinal number cards (1st – 8th)

1st	2nd
3rd	4th
5th	6th
7th	8th

New Heinemann Maths 3 • © SPMG 2000 • Copying permitted for purchasing school only. This material is not copyright free.

Ordinal number cards (9th – 16th)

9th	10th
11th	12th
13th	14th
15th	16th

Ordinal number cards (17th – 20th)/names (first – second)

NHM 3 — New Heinemann Maths — 43

| 17th | 18th |
| 19th | 20th |

first

second

New Heinemann Maths 3 • © SPMG 2000 • Copying permitted for purchasing school only. This material is not copyright free.

Ordinal number names (third – sixth)

third

fourth

fifth

sixth

(Ordinal number names (seventh – tenth)

seventh

eighth

ninth

tenth

Ordinal number names (eleventh – fourteenth)

eleventh

twelfth

thirteenth

fourteenth

Ordinal number names (fifteenth – eighteenth)

fifteenth

sixteenth

seventeenth

eighteenth

Ordinal number names (nineteenth/twentieth/fiftieth/hundredth)

48

nineteenth

twentieth

fiftieth

hundredth

Ten frames

Time cards (am ⟶ pm times)

12.00 midnight	1.00 am
2.00 am	3.00 am
4.00 am	5.00 am
6.00 am	7.00 am

Time cards (am ⟶ pm times)

8.00 am	9.00 am
10.00 am	11.00 am
12.00 noon	1.00 pm
2.00 pm	3.00 pm

Time cards (am ⟶ pm times)

4.00 pm	5.00 pm
6.00 pm	7.00 pm
8.00 pm	9.00 pm
10.00 pm	11.00 pm

Isometric dot grid